PEARSON

Math
Makes Sense

4

WNCP Edition

Peggy Morrow

Lorraine Kinsman

Ricki Wortzman

Sharon Jeroski

Ray Appel

Karen Campbell

Lalie Harcourt

Trevor Brown

With Contributions from

Daryl Chichak

Bryn Keyes

Ralph Connelly

Jason Johnston

Michael Davis

Angela D'Alessandro

Don Jones

Angie Harding

PEARSON
Education
Canada

Publisher
Claire Burnett

Publishing Team
Enid Haley
Lesley Haynes
Winnie Siu
Lynne Gulliver
Stephanie Cox
Cheri Westra
Judy Wilson

Photo Research
Karen Hunter

Design
Word & Image Design Studio Inc.

Math Team Leader
Diane Wyman

Product Manager
Kathleen Crosbie

ISBN-13: 978-0-321-43799-0
ISBN-10:   0-321-43799-3

Printed and bound in the United States

1 2 3 4 5 – 11 10 09 08 07

The information and activities presented in this
book have been carefully edited and reviewed.
However, the publisher shall not be liable for any
damages resulting, in whole or in part, from the
reader's use of this material.

Brand names that appear in photographs of
products in this textbook are intended to provide
students with a sense of the real-world applications
of mathematics and are in no way intended to
endorse specific products.

The publisher wishes to thank the staff and
students of Herchmer Elementary School for their
assistance with photography.

Consultants and Advisers

Consultants

Craig Featherstone
Mignonne Wood
Trevor Brown

Assessment Consultant
Sharon Jeroski

Elementary Mathematics Adviser
John A. Van de Walle

Cultural Adviser
Susan Beaudin

Advisers

Pearson Education thanks its Advisers, who helped shape the vision for *Pearson Math Makes Sense* through discussions and reviews of manuscript.

Joanne Adomeit
Bob Belcher
Bob Berglind
Auriana Burns
Steve Cairns
Edward Doolittle
Brenda Foster
Marc Garneau
Angie Harding
Florence Glanfield

Jodi Mackie
Ralph Mason
Christine Ottawa
Gretha Pallen
Shannon Sharp
Cheryl Shields
Gay Sul
Chris Van Bergeyk
Denise Vuignier

Reviewers

Field Test and Pilot Teachers

Pearson Education would like to thank the teachers and students who field-tested or piloted *Pearson Math Makes Sense 4* prior to publication. Their feedback and constructive recommendations have been most valuable in helping us to develop a quality mathematics program.

Aboriginal Content Reviewers

Glenda Bristow, First Nations, Métis and Inuit Coordinator, St. Paul Education Regional Div. No. 1
Patrick Loyer, Consultant, Aboriginal Education, Calgary Catholic

Grade 4 Reviewers

Terri L. Ahl
Regina Public S.B. #4, SK

Karen Anderson
Calgary Board of Education, AB

Ron Blackmer
Medicine Hat Catholic Board of Education, AB

Wendy Campbell
Calgary Board of Education, AB

Jacky Carter
Calgary Separate School District, AB

Kim Domoslai
Calgary Board of Education, AB

Fiona A. Eckert
School District #38 (Richmond), BC

Denise Flick
School District #20 (Kootenay-Columbia), BC

Gail Jenkins
Edmonton Catholic School Board, AB

Linda Judd
School District #23 (Central Okanagan), BC

Angela M. Kelly
Regina Public S.B. #4, SK

Christine Kennerd
Edmonton Catholic School Board, AB

Blaise Kirchgesner
St. Paul's RCSSD 20 (Saskatoon), SK

Robert Larson
Battle River School Division, AB

Jenny Miller
Calgary Board of Education, AB

Peter V. Miller
Battle River School Division #31, AB

Barb O'Connor
Calgary Board of Education, AB

Margaret Petruk
Edmonton Catholic School Board, AB

Cynthia Pratt Nicolson
University of British Columbia

Carole Saundry
School District #38 (Richmond), BC

Laura A.K. Schorn
Calgary Board of Education, AB

Tracy Smith
Elk Island Public Schools, AB

David Whiteside
Calgary Board of Education, AB

Table of Contents

Multiplication and Division Facts

Measurement

UNIT 7

Data Analysis

UNIT 8

Multiplying and Dividing Larger Numbers

Welcome to
Pearson Math Makes Sense 4

Math helps you understand what you see and do every day.

You will use this book to learn about the math around you. Here's how.

In each Unit:

- A scene from the world around you reminds you of some of the math you already know.

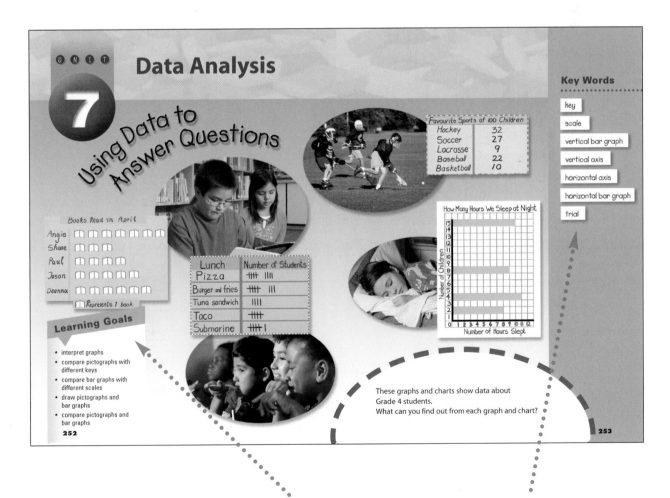

Find out what you will learn in the **Learning Goals** and important **Key Words**.

In each Lesson:

> You **Explore** an idea or problem, usually with a partner. You often use materials.

> Then you **Show and Share** your results with other students.

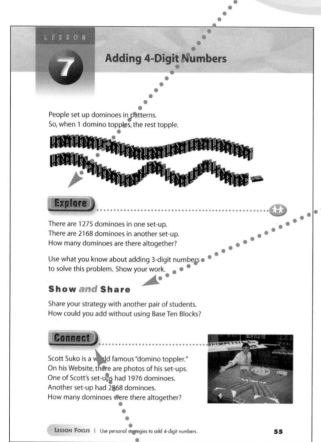

> **Connect** summarizes the math. It often shows a solution, or multiple solutions, to a question.

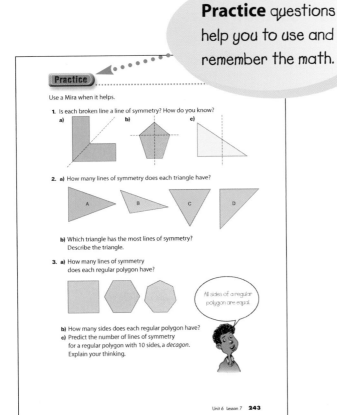

Practice

Use a Mira when it helps.

1. Is each broken line a line of symmetry? How do you know?
   a)   b)   c)

2. a) How many lines of symmetry does each triangle have?
   A   B   C   D

   b) Which triangle has the most lines of symmetry? Describe the triangle.

3. a) How many lines of symmetry does each regular polygon have?

   All sides of a regular polygon are equal.

   b) How many sides does each regular polygon have?
   c) Predict the number of lines of symmetry for a regular polygon with 10 sides, a *decagon*. Explain your thinking.

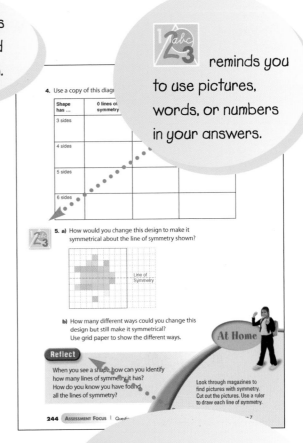

4. Use a copy of this diagram

Shape has ...	0 lines of symmetry
3 sides	
4 sides	
5 sides	
6 sides	

5. a) How would you change this design to make it symmetrical about the line of symmetry shown?

   Line of Symmetry

   b) How many different ways could you change this design but still make it symmetrical? Use grid paper to show the different ways.

At Home

Look through magazines to find pictures with symmetry. Cut out the pictures. Use a ruler to draw each line of symmetry.

Reflect

When you see a shape, how can you identify how many lines of symmetry it has? How do you know you have found all the lines of symmetry?

- Learn about strategies to help you solve problems in each **Strategies Toolkit** lesson.

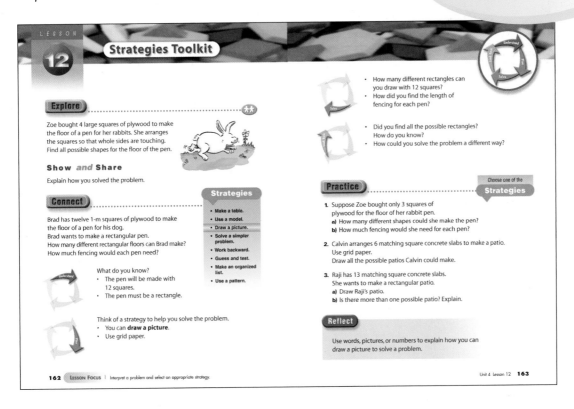

LESSON 12

Strategies Toolkit

Explore

Zoe bought 4 large squares of plywood to make the floor of a pen for her rabbits. She arranges the squares so that whole sides are touching. Find all possible shapes for the floor of the pen.

Show and Share

Explain how you solved the problem.

Connect

Brad has twelve 1-m squares of plywood to make the floor of a pen for his dog. Brad wants to make a rectangular pen. How many different rectangular floors can Brad make? How much fencing will each pen need?

What do you know?
- The pen will be made with 12 squares.
- The pen must be a rectangle.

Think of a strategy to help you solve the problem.
- You can **draw a picture**.
- Use grid paper.

Strategies
- Make a table.
- Use a model.
- Draw a picture.
- Solve a simpler problem.
- Work backward.
- Guess and test.
- Make an organized list.
- Use a pattern.

- How many different rectangles can you draw with 12 squares?
- How did you find the length of fencing for each pen?

- Did you find all the possible rectangles? How do you know?
- How could you solve the problem a different way?

Practice

Choose one of the **Strategies**

1. Suppose Zoe bought only 3 squares of plywood for the floor of her rabbit pen.
   a) How many different shapes could she make the pen?
   b) How much fencing would she need for each pen?

2. Calvin arranges 6 matching square concrete slabs to make a patio. Use grid paper. Draw all the possible patios Calvin could make.

3. Raji has 13 matching square concrete slabs. She wants to make a rectangular patio.
   a) Draw Raji's patio.
   b) Is there more than one possible patio? Explain.

Reflect

Use words, pictures, or numbers to explain how you can draw a picture to solve a problem.

- Check up on your learning in **Show What You Know** and **Cumulative Review**.

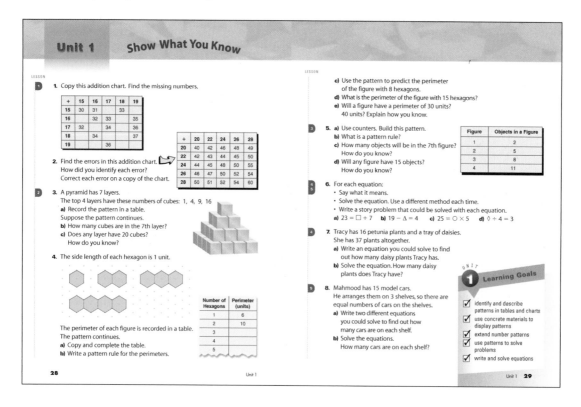

- The **Unit Problem** returns to the opening scene.

  It presents a problem to solve or a project to do using the math of the unit.

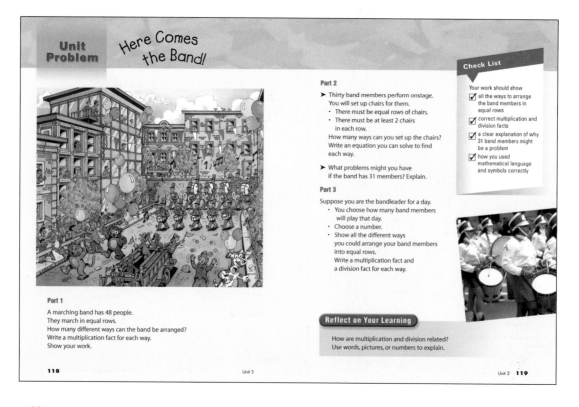

Explore some interesting math when you do the **Investigations**.

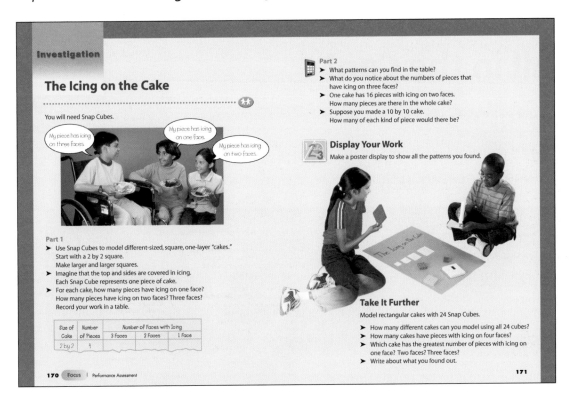

You will see **Games** pages.

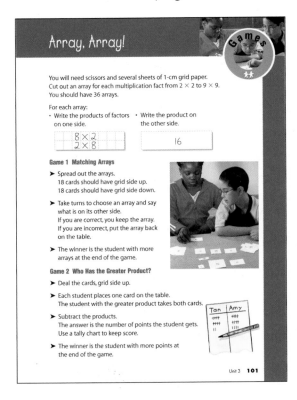

The **Glossary** is an illustrated dictionary of important math words.

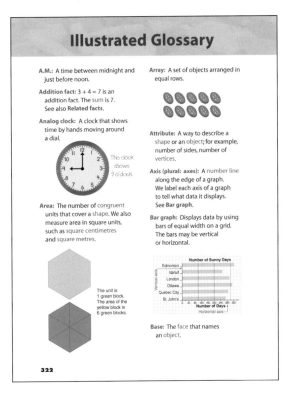

It's All in the Can!

You will need a variety of tin cans with labels
and a centimetre measuring tape.

Part 1

➤ Choose six cans. Label the cans with the letters from A to F.
   For each can, predict which measurement will be greater:
   the height of the can, or the distance around the can.
   Record your predictions.

➤ Measure the height and the
   distance around each can.
   Record the measurements to
   the nearest centimetre.
   Were your predictions correct?

➤ For each can, find the difference of the distance
around the can and its height.
Subtract the lesser number from the greater number.
➤ Record the measurements and the differences in a table.

Can	Description	Height	Distance Around	Difference
A	Spaghetti Sauce	14 cm	26 cm	26 – 14 = 12
B	Salmon	5 cm	25 cm	25 – 5 = 20

Part 2

➤ Arrange the cans in order from the
can with the least difference of the
two measurements to the can with
the greatest difference.
➤ Describe any patterns in the shapes
of the cans.
➤ Choose three different cans. Predict
their order from the can with the
least difference of the two
measurements to the can with the
greatest difference.
➤ Measure the cans to check your
prediction.

Display Your Work

Report your findings using words, pictures, or numbers.

Take It Further

Think of other ways you can sort the cans.
Arrange the cans in order from greatest to least.
Write about your sorting.

UNIT 1

Calendar Patterns

Patterns and

			January			
Sunday	**Monday**	**Tuesday**	**Wednesday**	**Thursday**	**Friday**	**Saturday**
	1	2	3	4	5	6
7	8	9	10	11	12	13
14	15	16	17	18	19	20
21	22	23	24	25	26	27
28	29	30	31			

Learning Goals

- identify and describe patterns in tables and charts
- use concrete materials to display patterns
- extend number patterns
- use patterns to solve problems
- write and solve equations

			April			
Sunday	**Monday**	**Tuesday**	**Wednesday**	**Thursday**	**Friday**	**Saturday**
1	2	3	4	5	6	7
8	9	10	11	12	13	14
15	16	17	18	19	20	21
22	23	24	25	26	27	28
29	30					

4

Equations

July

Sunday	Monday	Tuesday	Wednesday	Thursday	Friday	Saturday
1	**2**	**3**	**4**	**5**	**6**	**7**
8	**9**	**10**	**11**	**12**	**13**	**14**
15	**16**	**17**	**18**	**19**	**20**	**21**
22	**23**	**24**	**25**	**26**	**27**	**28**
29	**30**	**31**				

October

Sunday	Monday	Tuesday	Wednesday	Thursday	Friday	Saturday
	1	2	3	4	5	6
7	8	9	10	11	12	13
14	15	16	17	18	19	20
21	22	23	24	25	26	27
28	29	30	31			

Key Words

pattern rule

equation

solve an equation

solution

- What patterns do you see in these calendar pages?
- How might patterns change when the first day of the month is on Monday instead of Sunday?

5

Patterns in Charts

Look at this hundred chart.
There is a pattern in the numbers.
There is a pattern in the positions of the coloured squares.

Describe the patterns you see.

1	2	3	4	5	6	7	8	9	10
11	12	13	14	15	16	17	18	19	20
21	22	23	24	25	26	27	28	29	30
31	32	33	34	35	36	37	38	39	40
41	42	43	44	45	46	47	48	49	50
51	52	53	54	55	56	57	58	59	60
61	62	63	64	65	66	67	68	69	70
71	72	73	74	75	76	77	78	79	80
81	82	83	84	85	86	87	88	89	90
91	92	93	94	95	96	97	98	99	100

Explore

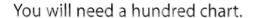

You will need a hundred chart.

➤ Decide on a number pattern.
   Keep it secret.
   Write the first 10 numbers in your pattern.
   Erase 3 numbers in your pattern.

➤ Trade patterns with your partner.
   Describe your partner's pattern.
   Identify the missing numbers.
   Extend the pattern.
   Write the next 4 numbers.

Show and Share

Talk with your partner.
How did you know how to extend your partner's pattern?
How did you find the missing numbers?

Here is the start of a pattern
on a hundred chart.

You can describe the pattern
in different ways.
These are **pattern rules**.

1	2	3	4	5	6	7	8	9	10
11	12	13	14	15	16	17	18	19	20
21	22	23	24	25	26	27	28	29	30
31	32	33	34	35	36	37	38	39	40
41	42	43	44	45	46	47	48	49	50
51	52	53	54	55	56	57	58	59	60
61	62	63	64	65	66	67	68	69	70
71	72	73	74	75	76	77	78	79	80
81	82	83	84	85	86	87	88	89	90
91	92	93	94	95	96	97	98	99	100

➤ Look at the positions of the
   coloured squares.
   Starting at 2, every third square
   is coloured.

 One pattern rule is:

Use 2 as the start diagonal.
The coloured squares lie along every third diagonal.
The diagonals go 1 down, 1 left.

➤ Look at the numbers in the coloured squares.
   The first 10 numbers in the pattern are:
   2, 5, 8, 11, 14, 17, 20, 23, 26, 29

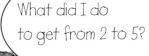

What did I do
to get from 2 to 5?

 Another pattern rule is:

Start at 2. Count on by 3s.

You can complete the pattern in the
chart using either rule above.

1	2	3	4	5	6	7	8	9	10
11	12	13	14	15	16	17	18	19	20
21	22	23	24	25	26	27	28	29	30
31	32	33	34	35	36	37	38	39	40
41	42	43	44	45	46	47	48	49	50
51	52	53	54	55	56	57	58	59	60
61	62	63	64	65	66	67	68	69	70
71	72	73	74	75	76	77	78	79	80
81	82	83	84	85	86	87	88	89	90
91	92	93	94	95	96	97	98	99	100

➤ Continue to colour the numbers
   that lie along the diagonals
   that go 1 down, 1 left.
   Colour new diagonals
   to continue the pattern.

➤ Continue to add 3.
   …, 29, 32, 35, 38, 41, 44, 47, 50, …

Practice

1. On the same hundred chart:
   - Start at 3. Count on by 3s to 100.
     Shade these numbers with one colour.
   - Start at 4. Count on by 4s to 100.
     Shade these numbers with another colour.
   a) Look at the numbers that are shaded in both colours.
      Describe the pattern in these numbers.
   b) What is a rule for this new pattern?

2. Anthony has guitar lessons every
   Wednesday in April.
   His sister has piano lessons
   every third day, starting April 2nd.
   a) On what date do both Anthony
      and his sister have a lesson?
   b) How did you solve the problem?

April						
S	M	T	W	T	F	S
	1	2	3	4	5	6
7	8	9	10	11	12	13
14	15	16	17	18	19	20
21	22	23	24	25	26	27
28	29	30				

3. Find the missing numbers in this
   hundred chart.
   What strategies did you use?

	102	103	104	105	106	107	108	109	110
111	112	113	114	115			118	119	120
121	122	123	124	125			128	129	130
131	132	133	134		136	137	138	139	140
141			144				148	149	150
151	152	153	154	155	156	157	158	159	160
161	162	163	164	165	166		168	169	170
171	172		174	175	176	177	178	179	180
181	182	183					188	189	190
191	192	193	194	195	196	197	198	199	200

4. On 1-cm grid paper, make a
   5-wide hundred chart with
   5 columns and 20 rows.
   a) Find five different patterns in
      the 5-wide hundred chart.
      Record the patterns.
   b) How do the patterns in a
      5-wide hundred chart
      compare to the patterns in a
      10-wide hundred chart?
   Show your work.

5-Wide Hundred Chart

1	2	3	4	5
6	7	8	9	10
11	12	13	14	15

5. Explain how these two patterns are related.

×	1	2	3	4	5
1	1	2	3	4	5
2	2	4	6	8	10
3	3	6	9	12	15
4	4	8	12	16	20
5	5	10	15	20	25

1	2	3	4	5
6	7	8	9	10
11	12	13	14	15
16	17	18	19	20
21	22	23	24	25

6. Identify the errors in this addition chart.
How did you identify each error?
Correct each error.

+	22	24	26	28	30
11	33	35	37	39	40
13	35	36	39	41	43
15	37	39	40	43	45
17	38	41	43	45	47
19	41	42	45	46	49

7. Look at the coloured squares in this
addition chart.

+	10	11	12	13	14	15
10	20	21	22	23	24	25
11	21	22	23	24	25	26
12	22	23	24	25	26	27
13	23	24	25	26	27	28
14	24	25	26	27	28	29
15	25	26	27	28	29	30

a) Describe the pattern in two ways.
b) Write a pattern rule for the number pattern.

At Home

Reflect

Sometimes it is difficult to
find a pattern rule.
What can you do
if you are stuck?

What number patterns do you
see at home?
Look through magazines and
newspapers.
Cut out any patterns you find.

Extending Number Patterns

Explore

You will need a geoboard, geobands, and dot paper.

➤ Use the geoboard to make a rectangle with length 2 units and width 1 unit. Count and record the number of pegs on the perimeter of the rectangle.

➤ Make a rectangle with length 3 units and width 2 units. Count and record the number of pegs on the perimeter.

➤ Continue to make rectangles with length 1 unit greater than the width. Record the length, the width, and the number of pegs each time.

Draw each rectangle on dot paper.

Rectangle	Length	Width	Number of Pegs on Perimeter
1	2	1	6
2	3	2	

• How many pegs will be on the perimeter of the 5th rectangle? The 8th rectangle? How do you know?
• Will the perimeter of any rectangle have 32 pegs? 33 pegs? 34 pegs? How do you know?

Show *and* Share

Share your results with another pair of classmates.
What patterns do you see in the table?
How did you use these patterns to solve the problems?

Here is a pattern of triangles drawn on dot paper.
Each triangle has 2 equal sides.

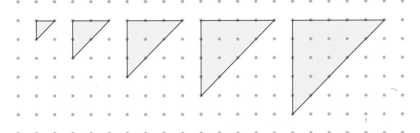

Count the dots
on each
perimeter.

This pattern continues.

➤ Find the number of dots on the perimeter of the
12th triangle.
Make a table.

Triangle Number	Number of Dots on Perimeter
1	3
2	6
3	9
4	12
5	15

Skip count by 3 to
extend the pattern.

One pattern rule for the number
of dots on the perimeter is:

Start at 3. Add 3 each time.

Another pattern rule is:

The triangle number
multiplied by 3

For the 12th triangle, skip count by 3 twelve times:
3, 6, 9, 12, 15, 18, 21, 24, 27, 30, 33, 36
The 12th triangle will have 36 dots on its perimeter.

➤ Will any triangle have 22 dots on its perimeter?
The number of dots on any perimeter is a number we get
when we start at 3 and skip count by 3.
Since 22 is not one of those numbers, a triangle in this pattern
cannot have 22 dots on its perimeter.

1. Here is a pattern of figures made with Colour Tiles.

Figure 1   Figure 2   Figure 3   Figure 4

The pattern continues.

a) Draw the next two figures on grid paper.

b) Copy and complete the table for the first 6 figures.

Figure	Number of Green Tiles	Number of Yellow Tiles
1	2	10

c) Write a pattern rule for the number of green tiles.

d) Write a pattern rule for the number of yellow tiles.

e) How many green tiles will be in the 8th figure?

f) How many yellow tiles will be in the 10th figure?

g) Will any figure have 21 green tiles? 31 yellow tiles?
   Describe how you made your decision.

2. Regular pentagons are combined to make new figures.
   Each pentagon touches no more than 2 other pentagons.

   The side length of each pentagon is 1 unit.
   The perimeter of each figure is recorded in a table.

Number of Pentagons	Perimeter (units)
1	5
2	8

Figure 1

Figure 2

Figure 3

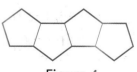

Figure 4

a) Copy and complete the table for the first 4 figures.

b) Write a pattern rule for the perimeters.

c) Use the pattern to predict the perimeter
   of the figure with 6 pentagons.
   With 10 pentagons.

3. Pizza Parlour has prices that follow a pattern.

Pizza Number	Pizza Size	Price with Cheese ($)	Price with Two Toppings ($)
1	4-slice	5	7
2	8-slice	10	12
3	12-slice	15	17
4	16-slice	20	
5	20-slice		

a) Copy and complete the table.
b) Write a pattern rule for the price with cheese.
c) Write a pattern rule for the price with two toppings.
d) Suppose the patterns in the table continue.
What is the price of a 48-slice pizza with two toppings?
e) How is the price of the pizza with two toppings related to the price of the pizza with cheese?

4. Each package of cards contains 5 cartoon cards.
a) Copy and complete this table for the first 5 packages.

Number of Packages	Number of Cards
1	
2	

b) Write a pattern rule for the number of cards.
c) Find the number of cards in 9 packages and in 15 packages.
d) The deluxe edition contains 4 packages in a tin.
How many cards will there be in 3 deluxe tins? 7 deluxe tins?
Show your work.

Reflect

How can a table help you solve a problem?
Use an example to show your thinking.

Music

There are many patterns in music. A melodic ostinato is a short pattern in the melody. It repeats throughout a song.

Representing Patterns

You will need congruent Pattern Blocks and grid paper or dot paper.

Figure	Number of Blocks in a Figure
1	2
2	4
3	6
4	8
5	10
6	12

➤ Build the first 6 figures of this pattern. Make sure the figures show a pattern.

➤ Draw your pattern on grid paper or dot paper.

➤ Use your model of the pattern or the table. Build, then draw the next 3 figures in your pattern.

➤ How many blocks would you need for the 12th figure in your pattern? How do you know?

Show *and* Share

Compare your pattern with that of another pair of classmates.
If the patterns are different, is one pattern incorrect? Explain.
Work together to write a pattern rule for the number of blocks in a figure.
How many blocks would you need for the 15th figure?
Build the figure to check.

Here is a pattern.

Figure	Counters in a Figure
1	1
2	3
3	5
4	7

+2
+2
+2

From the table, the Counters in a Figure increase by 2.

We can use counters to build this pattern in different ways:
Pattern 1

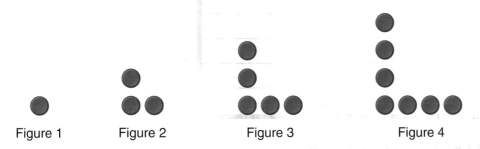

Figure 1   Figure 2   Figure 3   Figure 4

Each figure has 2 more counters than the figure before.
Pattern 2

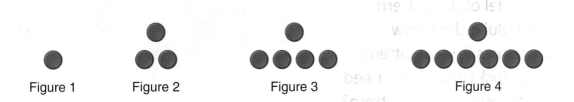

Figure 1   Figure 2   Figure 3   Figure 4

Each figure has 2 more counters than the figure before.
From the table or from the counters, the pattern rule for the
number of counters in a figure is: Start at 1. Add 2 each time.

To find the number of counters in Figure 10, start at Figure 4 with
7 counters and skip count by 2, six times: 7, 9, 11, 13, 15, 17, 19
There will be 19 counters in Figure 10.

Here is Figure 10 for Pattern 1:

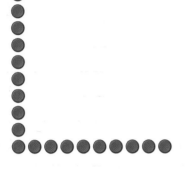

Here is Figure 10 for Pattern 2:

Practice

1. **a)** Use counters to build this pattern.

Figure	Counters in a Figure
1	4
2	8
3	12
4	16

   b) What is a pattern rule?

2. **a)** Use Pattern Blocks to build this pattern.
   b) Compare your pattern with that of a classmate who used different Pattern Blocks. How can you tell the patterns have the same rule?
   c) Write a pattern rule.
   d) How many blocks would be in the 9th figure? How did you find out?

Figure	Blocks in a Figure
1	4
2	7
3	10
4	13

3. a) Use congruent squares.
   Build this pattern.
   Record the pattern on grid paper.
   b) Find the missing data in the table.
   How can you check that your answers are correct?

Figure	Squares in a Figure
1	6
2	8
3	10
4	
5	14
6	

4. Joe made a design with 8 strips of coloured tape.
   He recorded the first 4 strip lengths in a table.
   a) Draw the 8 strips on 1-cm grid paper.
   Record each length.
   b) Suppose the pattern continues.
   What is a pattern rule?
   c) Predict the length of the 10th strip.

Strip	Length
1	1 cm
2	2 cm
3	4 cm
4	7 cm

5. Nicole made up a pattern.
   She recorded some of her pattern in this table.
   a) What might Nicole's pattern look like?
   Use Pattern Blocks.
   Build as many different patterns as you can.
   Record each pattern on square or triangular dot paper.
   b) For each pattern you build, draw the 7th figure.

Figure	Blocks in a Figure
1	1
2	
3	
4	10

Reflect

When you show a pattern two ways, how can you check they match?
Use a pattern you have built to explain.

Equations Involving Addition and Subtraction

An **equation** is a statement that two things are equal.

We can use counters to show an addition equation:

We write: $5 + 3 = 8$

We can use counters to show a subtraction equation:

We write: $10 - 4 = 6$

Explore

You will need equation cards, blank cards, scissors, and Base Ten Blocks.
Cut out all the cards.
Place them so you can read them.

To play the game, match each equation with its story,
then use the Base Ten Blocks to show each equation.
Score 3 points for a correct equation, story,
and Base Ten Block picture.
Take turns until all the cards have been used.
Use the blank cards to write a story and its equation.
Use the blocks to show the equation for your story.

Master 1.16   Card Sort

1. Ramjit has completed 7 levels in a video game. The game has 29 levels altogether. How many more levels does Ramjit have to play to complete all levels?	2. A recreation park has a dozen sections. The city plans to close some sections to reduce the number to 7. How many sections will be closed?	3. Janet's friend gave her 12 sports cards. Janet now has 29 cards. How many cards did she begin with?
4. It's a leap year! Aiden's birthday is on March 7. He starts counting days from February 1. How many days will he count until his birthday?	5. The school newsletter publishes 12 articles every month. This month, 7 of the articles submitted were turned down. How many articles were submitted?	6. Jetta caught 29 fish. After giving some away, she had 12 fish left. How many fish did Jetta give away?
A. $12 - \square = 7$	B. $7 + \triangle = 29$	C. $12 = \bigcirc - 7$
D. $12 = 29 - \blacktriangle$	E. $29 = \bullet + 12$	F. $29 + 7 = \blacksquare$

The right to reproduce or modify this page is restricted to purchasing schools.
This page may have been modified from its original. Copyright © 2007 Pearson Education Canada Inc.

Show *and* Share

What strategies did you use to find a match?
Share the matches with another pair of classmates.
Have your classmates check the equation you wrote for your story.

Les and Rae collect rocks. Rae has 12 rocks.
Together, they have a total of 32 rocks.
Write an equation to represent
how many rocks Les has.

We use a symbol to represent the number
of rocks Les has.

You can use any symbol you like for the unknown number. We use □.

Let □ represent the number of rocks Les has.
We know that:
Les' rocks + Rae's rocks = 32
Rae has 12 rocks.
So, we can write this equation: □ + 12 = 32

*To **solve an equation** means to find the value of the unknown number.*

Here are 3 possible ways to solve this equation.

- Use counters.
  The total number of rocks is 32.
  Rae has 12 rocks.
  Use 32 counters.
  Divide the counters into 2 groups.
  One group has 12 counters.
  The other group has 20 counters.
  These are Les' rocks.

Rae's rocks  Les' rocks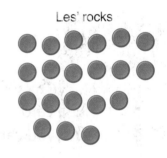

- Draw a picture.
  Les has 20 rocks.

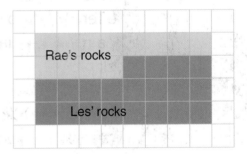

Rae's rocks

Les' rocks

- Use guess and test, and mental math.
  □ + 12 = 32
  Guess a number for □, then test to see if you are correct.
  Guess: □ = 10
  Test: 10 + 12 = 22   This is too low.
  Guess: □ = 15
  Test: 15 + 12 = 27   This is too low, but closer to the number we want.

Guess: ☐ = 20
Test: 20 + 12 = 32   This is correct.

Les has 20 rocks.

> ☐ = 20 is the **solution** to the equation.

Practice

1. Write an equation for each set of Base Ten Blocks.
   a)

   b)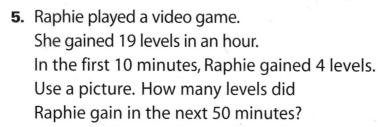

2. Say what each equation means.
   Use counters to solve each equation.
   a) ☐ + 5 = 11   b) 3 + ☐ = 15   c) 12 = ☐ + 1   d) 14 = 3 + ☐

3. Say what each equation means.
   Use counters to solve each equation.
   a) ∇ − 4 = 9   b) 13 − ∇ = 10   c) 17 = ∇ − 2   d) 21 = 27 − ∇

For questions 4 to 6:
Write an equation that represents the question.
Solve the equation using the method given.

4. Melissa and Tyler have music folders on
   their personal music players.
   Altogether, they have a total of 16 folders.
   Melissa has 4 folders. Use counters.
   How many folders does Tyler have?

5. Raphie played a video game.
   She gained 19 levels in an hour.
   In the first 10 minutes, Raphie gained 4 levels.
   Use a picture. How many levels did
   Raphie gain in the next 50 minutes?

6. Mandeep had a hole in his pocket.
   He started with 79 cents in his pocket.
   On the way home, 23 cents fell out. Use guess and test.
   How much money did Mandeep have in his pocket when he got home?
   Record your guesses.

7. The Sidhu family went on vacation.
   The family drove 213 km the 1st day,
   122 km the 2nd day, rested on
   the 3rd day, and drove a long distance
   on the 4th day. The family drove
   a total distance of 763 km.
   Which equation represents how
   far the family drove on the 4th day?
   Explain your choice.

   a) $213 + \nabla + 122 = \square$ b) $213 - 122 + \nabla = 763$
   c) $213 + 122 + 0 = 763$ d) $213 + 122 + \nabla = 763$

8. a) Write a story problem you could solve using this equation: $3 + \square = 11$
   b) Solve the equation.
   c) What is the answer to the problem?

9. a) Write a story problem you could solve using this equation: $30 = 34 - \square$
   b) Solve the equation.
   c) What is the answer to the problem?

10. Use these numbers and some of the symbols: $4, 15, \square, +, -, =$
    a) Write an equation.
       How many different equations can you write?
    b) Solve each equation. Use a different method each time.
    c) Write a story problem for each equation.
       Use your answers in part b to solve each problem.
    Show your work.

Reflect

Talk to a partner.
Tell how you choose the method you use to solve an equation.

Equations Involving Multiplication and Division

Explore ·  Game

 You will need a calculator, 66 counters, two number cubes labelled 1 to 6, and blank squares.
On 4 of the blank squares, draw: ×, ÷, =, □

Player A rolls the number cubes.
Each number rolled is either a tens digit or a ones digit.
For example, a 2 and a 4 could be 24 or 42.
Player A writes the 2-digit number on a blank square.
Player A writes a factor of the 2-digit number on a blank square. She can use a calculator if necessary.
For example, a factor of 24 is 6.
Player A then uses these cards to make a multiplication equation, such as:

The symbol □ represents the unknown number in the equation.

Player B has to use counters to find the unknown number.
If the question can be done, but Player B cannot do it,
Player A gets a point.
If the question is impossible, then Player B gets a point.
Record all your equations.
Take turns until one player gets 5 points.

Show *and* Share

Share your equations with another pair of classmates.
Compare strategies for finding the unknown number.

Sarah has 20 apples. She shares them among 5 friends.
Write an equation to represent
how many apples each friend gets.

We can use a symbol to represent this number.
Let ◊ represent the number of apples each friend gets.
We know that:

| the total number of apples | ÷ | the number of apples each friend gets | = | the numbers of friends |

There are 20 apples in total. There are 5 friends.
So, we can write the equation: $20 \div ◊ = 5$

Here are 3 ways to solve this equation.

- Use counters.
  The total number of apples is 20.
  There are 5 friends.
  Use 20 counters.
  Divide the counters into 5 equal groups.

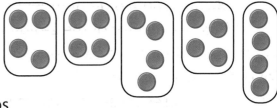

  Each group has 4 counters.
  So, each friend gets 4 apples.

- Draw a picture.
  Use grid paper.
  Draw an array of 20 squares,
  with 5 squares in each row.

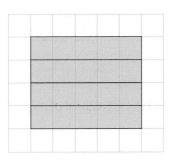

  There are 4 rows.
  So, each friend gets 4 apples.

- Use mental math.
  $20 \div ◊ = 5$
  Think of a related multiplication fact.
  What do we multiply 5 by to get 20?
  $5 \times 4 = 20$
  So, $20 \div 4 = 5$
  And, $◊ = 4$
  Each friend gets 4 apples.

◊ = 4 is the solution to the equation.

1. Write a multiplication equation for each array.

   a)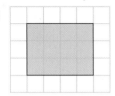

   b)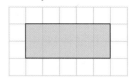

2. Write a division equation for each array in question 1.

3. Say what each equation means.
   Use counters to solve each equation.

   a) $\triangle \times 3 = 9$ **b)** $5 \times 2 = \square$ **c)** $\bigcirc \div 2 = 4$ **d)** $5 = \Diamond \div 1$

4. Write a story problem that could be solved using each equation in question 3.

5. Sholeh, Mark, Tasha, and Cedar practised relay around the track.
   Sholeh ran 2 laps, then passed the baton to Cedar.
   Cedar ran 2 laps, then passed the baton to Mark.
   Mark ran 2 laps, then passed the baton to Tasha who ran 2 laps.
   Which equation could you use to find how many laps the students ran altogether? Explain your choice.

   a) $8 \div 2 = \square$ **b)** $4 \times \square = 8$ **c)** $2 \times 4 = \square$ **d)** $2 \times 8 = \square$

6. Use these numbers and some of the symbols: $2, 6, \square, \times, \div, =$
   a) Write an equation. How many different equations can you write?
   b) Solve each equation. Use a different method each time.
   c) Write a story problem for each equation.
      Use your answers to part b to solve each problem.
   Show your work.

7. Salim has 7 friends. Each friend has 12 books.
   a) Write an equation to represent how many
      books Salim's friends have altogether.
   b) Solve the equation. Solve the problem.

Reflect

How can you check that your solution to an equation is correct?
Use an example to show your thinking.

Number the Blocks

You will each need 4 of each of these Pattern Blocks.

6 points   4 points   4 points   3 points

Each block has the number of points shown.

➤ Place another yellow block on the table.
➤ Take turns to place one of your blocks so it touches
  one side of the block on the table.

You can do this.   You *cannot* do this.

➤ Your score is the sum of the points for the block you placed
  and the block or blocks your block touches.

  For example,
  if you place the blue block,
  your score is 6 + 4 + 4 = 14.

➤ Continue playing until both players
  have no blocks left.

➤ The winner is the player with
  more points.

Strategies Toolkit

Explore

These equations have shapes in place of numbers.
Each shape represents a different number.
All the triangles represent the same number.
All the squares represent the same number.
All the circles represent the same number.

$$\blacksquare + \blacktriangle + \blacktriangle + \bullet = 17$$
$$\blacksquare + \blacktriangle + \bullet = 11$$
$$\blacksquare + \blacksquare + \blacktriangle = 8$$

Find the number that each shape represents.

Show and Share

Share the strategy you used to solve the problem.

Connect

Each shape represents a number.
$$14 = \heartsuit + \heartsuit + \blacktriangle + \blacktriangle$$
$$12 = \heartsuit + \heartsuit + \blacktriangle$$
$$10 = \bullet + \heartsuit + \blacktriangle$$

Find the number that each shape represents.

Understand

What do you know?
- Each shape represents a number.
- All the hearts represent one number.
- All the triangles represent one number.
- The circle represents one number.

Plan

Think of a strategy to help you solve the problem.
- You can **guess and test**.
- Guess a number for each shape.
  Test that the numbers fit the equations.

Strategies

- Make a table.
- Use a model.
- Draw a picture.
- Solve a simpler problem.
- Work backward.
- Guess and test.
- Make an organized list.
- Use a pattern.

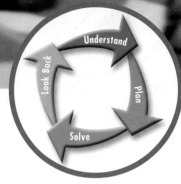

What are good guesses for ♥ and for ▲?
If the first two sums are not 14 and 12,
think about your next guesses.
Should each number be greater than or
less than your first guess?

Check your work.
Are the sums 14, 12, and 10?
How could you solve this problem another way?

Practice

Choose one of the

Strategies

1. Each shape represents a different number.
   Find the number that each shape represents.

   ■ + ■ + ● = 26

   ■ + ● + ● + ▲ = 24

   ■ + ● + ▲ = 18

2. Each letter represents a different number.
   Find the number that each letter represents.

   12 = A + B + C

   14 = A + A + B + B

   11 = A + B + B

3. Which object has the greatest mass?
   The least mass? Show your work.

Reflect

Persistence means sticking with something and not giving up.
Tell how persistence helped you solve these problems.

LESSON

1

1. Copy this addition chart. Find the missing numbers.

+	15	16	17	18	19
15	30	31		33	
16		32	33		35
17	32		34		36
18		34			37
19			36		

2. Find the errors in this addition chart.
How did you identify each error?
Correct each error on a copy of the chart.

+	20	22	24	26	28
20	40	42	46	48	49
22	42	43	44	45	50
24	44	45	48	50	55
26	46	47	50	52	54
28	50	51	52	54	60

2

3. A pyramid has 7 layers.
The top 4 layers have these numbers of cubes: 1, 4, 9, 16
a) Record the pattern in a table.
Suppose the pattern continues.
b) How many cubes are in the 7th layer?
c) Does any layer have 20 cubes?
How do you know?

4. The side length of each hexagon is 1 unit.

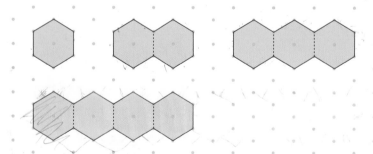

The perimeter of each figure is recorded in a table.
The pattern continues.
a) Copy and complete the table.
b) Write a pattern rule for the perimeters.

Number of Hexagons	Perimeter (units)
1	6
2	10
3	
4	
5	

c) Use the pattern to predict the perimeter of the figure with 8 hexagons.

d) What is the perimeter of the figure with 15 hexagons?

e) Will a figure have a perimeter of 30 units? 40 units? Explain how you know.

5. a) Use counters. Build this pattern.

b) What is a pattern rule?

c) How many objects will be in the 7th figure? How do you know?

d) Will any figure have 15 objects? How do you know?

Figure	Objects in a Figure
1	2
2	5
3	8
4	11

6. For each equation:
- Say what it means.
- Solve the equation. Use a different method each time.
- Write a story problem that could be solved with each equation.

a) $23 = \square + 7$ **b)** $19 - \triangle = 4$ **c)** $25 = \bigcirc \times 5$ **d)** $\lozenge \div 4 = 3$

7. Tracy has 16 petunia plants and a tray of daisies. She has 37 plants altogether.

a) Write an equation you could solve to find out how many daisy plants Tracy has.

b) Solve the equation. How many daisy plants does Tracy have?

8. Mahmood has 15 model cars. He arranges them on 3 shelves, so there are equal numbers of cars on the shelves.

a) Write two different equations you could solve to find out how many cars are on each shelf.

b) Solve the equations. How many cars are on each shelf?

UNIT

1 Learning Goals

- ✓ identify and describe patterns in tables and charts
- ✓ use concrete materials to display patterns
- ✓ extend number patterns
- ✓ use patterns to solve problems
- ✓ write and solve equations

Calendar Patterns

9 plus 17 is 26.

Part 1

Look at any 2 by 2 grid
on a calendar.
Add the pairs of numbers in
diagonally opposite corners.
What do you notice about the sums?
Is this true for all 2 by 2 grids?
Describe a rule for the pattern.
Explain why your rule makes sense.

What patterns can you find in a 3 by 3 grid?
A 4 by 4 grid?

24 plus 12 is 36.

Part 2

Try subtracting instead of adding.
Use different sizes of grids.
Describe any patterns.

Part 3

Write an equation using the number patterns
on a calendar.
Replace one number in your equation with a symbol.
Trade equations with a classmate.
Solve your classmate's equation.
Explain your method.

26 minus 10 is 16.

Reflect on Your Learning

Describe 2 things you learned about patterns in tables.
What have you learned about writing and solving equations?
Which Learning Goal was easiest for you?
Which was most difficult?

Whole Numbers

Those Amazing Elephants

Learning Goals

- recognize and read numbers from 1 to 10 000
- read and write numbers in standard form, expanded form, and written form
- compare and order numbers
- use diagrams to show relationships
- estimate sums and differences
- add and subtract 3-digit and 4-digit numbers mentally
- use personal strategies to add and subtract
- pose and solve problems

The elephant is the world's largest animal.
There are two kinds of elephants.
The African elephant can be found
in most parts of Africa.
The Asian elephant can be found in Southeast Asia.
African elephants are larger and heavier
than their Asian cousins.
The mass of a typical adult African
female elephant is about 3600 kg.
The mass of a typical male is about 5500 kg.
The mass of a typical adult Asian female
elephant is about 2720 kg.
The mass of a typical male is about 4990 kg.

- How could you find how much greater the mass of the
  African female elephant is than the Asian female elephant?
- Kandula, a male Asian elephant, had a mass of about 145 kg at birth.
  Estimate how much mass he will gain from birth to adulthood.
- The largest elephant on record was an African male with an
  estimated mass of about 10 000 kg. About how much greater was
  the mass of this elephant than the typical African male elephant?

Whole Numbers to 10 000

The largest marching band ever assembled had 4526 members.
There were students from 52 different school bands.

Explore

How many different ways can you show 4526?
Draw a picture to record each way you find.

Show and Share

Share your pictures with another pair of students.
How do you know each picture shows 4526?

Connect

The largest marching band had 1342 majorettes, flag bearers, and drill team members.
The rest of the band were musicians.
You can represent the number 1342 in different ways.

➤ Use Base Ten Blocks.
   To show the number 1342:

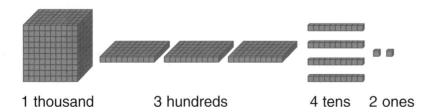

| 1 thousand | 3 hundreds | 4 tens | 2 ones |

➤ Draw a picture.
To show the number 1342:

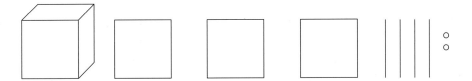

➤ Use a place-value chart.
To show the number 1342:

Thousands	Hundreds	Tens	Ones
1	3	4	2

1000   300   40   2

Every digit has a place value, depending on its position.

➤ Write the number 1342 as the sum of
the thousands, hundreds, tens, and ones.

1342 = 1000 + 300 + 40 + 2 ⟵———— This is **expanded form**.

➤ Use words.
1342 is one thousand three hundred forty-two.

➤ Use standard form.
The number 1342 is written in **standard form**.
It has no spaces between the digits.
The number 10 000 is also written in standard form.
        ↑

It has a space between the thousands digit and
the hundreds digit.

We do not use the word "and" when we represent whole numbers with words.

1. In 2001, the population of Iqaluit, Nunavut, was 5236.
   Write this number in words.

2. Mount Everest is the world's highest mountain.
   It is about 8850 m high.
   Use Base Ten Blocks to show this number.
   Draw pictures of the blocks.

3. Mount Logan, Yukon, is the highest mountain in Canada.
   It is about 5959 m high.
   Use expanded form to show this number.

4. Write the standard form of the number represented by
   each set of blocks.

   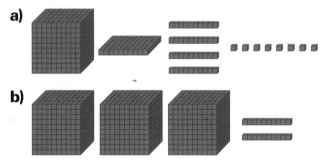

   a)

   b)

5. Write each number in question 4 in words.

6. Write each number in standard form.
   a) 5000 + 600 + 40 + 3   b) 9000 + 700 + 80
   c) 3000 + 200 + 9   d) 8000 + 20
   e) 7000 + 5   f) 4000 + 70 + 3

7. Write each number in standard form, then in expanded form.
   a) one thousand seven hundred fifty-four
   b) nine thousand nine hundred ninety-nine
   c) four thousand seventy
   d) six thousand five hundred three
   e) ten thousand

8. Write each number in expanded form.
   a) 5352   b) 7056   c) 8104   d) 4370

9. Draw a picture to represent each number in question 8.

10. Write each number in question 8 in words.

11. a) Press: 578. Make the screen show 508.
How did you do it?

b) Explain how to get each target number from each start number.

Start	394	156	4689
Target	94	106	4009

c) Write your own start number and target number.
Explain how you reached your target.

12. Tyler wrote 2005 in words as "two hundred five."
Explain Tyler's mistake.
Show your work.

13. Use Base Ten Blocks.
Find as many ways as you can to show 2058.
Record your work in a place-value chart.

14. Use a place-value chart to show each number.
a) 7649 **b)** 908 **c)** 9441 **d)** 39

15. The value of the 4 in 2413 is 400.
Write the value of each underlined digit.
a) 7<u>8</u>47 **b)** <u>9</u>305 **c)** 68<u>4</u>2 **d)** 9<u>9</u>99

16. A student read 7647 as
"seven thousand six hundred and forty-seven."
Explain the student's mistake.

At Home

Reflect

Use numbers, words, or pictures to explain the meaning of each digit in the number 7777.

Look through newspapers and magazines.
Find example of large numbers.
Write each number.
In which form is it written?

2

Comparing and Ordering Numbers

Explore

Use the digits 3, 5, 7, 8.
Write 3 different numbers using *all* these digits.
Order the numbers from greatest to least.
Show your work.

Show *and* Share

Share your numbers and ordering with another pair of students.
Take turns to tell about the strategy you used for ordering.
What other strategies could you use to order the numbers?

Connect

To order the numbers 2143, 2413, and 1423 from least to greatest:

➤ Represent each number with Base Ten Blocks.

2143

2413

Both 2143 and 2413
have 2 thousands.
Compare their hundred flats.
2143 has fewer.
So, 2143 is less than 2413.

1423

Compare the thousand cubes.
1423 has the fewest.
So, 1423 is the least number.

From least to greatest: 1423, 2143, 2413

LESSON FOCUS | Use different methods to compare and order numbers.

➤ Write each number in a place-value chart.

Thousands	Hundreds	Tens	Ones
2	1	4	3
2	4	1	3
1	4	2	3

1423 has the fewest thousands, so it is the least number.

Both 2143 and 2413 have 2 thousands.
Compare their hundreds.
100 is less than 400.
So, 2143 is less than 2413.

The arrow points to the smaller number.

You can use < and > to show order.
1423 < 2143 means
1423 is less than 2143.

2413 > 2143 means
2413 is greater than 2143.

5 > 2

➤ Use a number line.
Mark a dot for each number on a number line.

Read the numbers from left to right.
From least to greatest: 1423, 2143, 2413

Practice

1. The Canadian Armed Forces have 80 F-18 Hornets.
The US Navy has 200.
Which has more F-18s? How do you know?

2. Copy and complete. Write >, <, or =.
   a) 582 ☐ 589   b) 3576 ☐ 3476   c) 5754 ☐ 5745
   d) 792 ☐ 6082   e) 4110 ☐ 4101   f) 8192 ☐ 8291
   How did you decide which symbol to use?

3. Write the numbers in order from least to greatest.
   a) 862, 802, 869   b) 7656, 7665, 6756

4. Write the numbers in order from greatest to least.
   Explain how you did it.
   a) 9006, 9600, 9060   b) 5865, 895, 5685

5. Replace each ☐ with a digit so the statement is true.
   Write the possible digits for each ☐.
   a) 5762 < 5 ☐ 76   b) 7998 > ☐ 998   c) 6 ☐ 05 < 6604

6. Chantelle and Elena collect shells.
   Chantelle has 4325 shells.
   Elena has 4235.
   Who has more shells?
   How do you know?

7. Katie, Urvi, and Blake collect stamps.
   Katie has 2340 stamps.
   Urvi has 2304 stamps.
   Blake has 2430 stamps.
   Who has the most stamps? The fewest stamps?
   How do you know?

8. Write three 4-digit numbers.
   Order the numbers from greatest to least.

9. Use the digits 3, 7, 8, 9.
Write all the 4-digit numbers greater than 7000
and less than 8000.
Order the numbers from least to greatest.
Show your work.

10. Copy and fill in the blanks.
a) 8448, 8449, _____, _____, 8452, _____, _____
b) 5097, 5098, _____, _____, _____, 5102, _____
c) 4701, _____, _____, 4704, _____, _____, 4707
d) _____, 6320, _____, 6322, _____, _____, 6325

11. Rewrite the numbers in the correct order from
least to greatest.
a) 5228, 5229, 5231, 5232, 5230, 5233
b) 1009, 1014, 1012, 1015, 1010, 1013, 1011
c) 4438, 4440, 4439, 4441, 4443, 4442, 4437

12. Write the number for each letter on the number lines.

a)

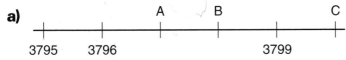

b)

c)

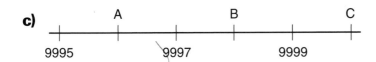

Reflect

Sue says that since 9 > 2, then 987 > 2134.
Is she correct?
Use words, pictures, or numbers to explain.

3

Sorting Numbers

Explore

You will need loops of string and numeral cards like those below.

224	3689	2313	1722	467	94

371	176	2388	4585	690	2000

➤ Sort the numbers using two attributes.
Record your sorting.

➤ Sort the numbers a different way.
Record the sorting.

Show and Share

Show another pair of classmates one way
you sorted. Ask them to tell the attributes
you used. Have them name one more
number for each of your groups.

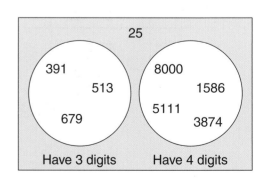

Connect

Here are four ways to sort these numbers.

8000   1586   391   5111   3874   513   679   25

➤ Use a **Venn diagram** with separate circles.

A number cannot have 3 digits *and* 4 digits,
so separate circles must be used.

Twenty-five has only 2 digits,
so it is outside the circles.

25

391		8000
	513	1586
		5111
679		3874

Have 3 digits   Have 4 digits

➤ Use a Venn diagram with one circle inside another circle.

All the numbers in the two circles are less than 5000.
The numbers in the inside circle also have 4 digits.

Both 8000 and 5111 are greater than 5000, so they are outside the circles.

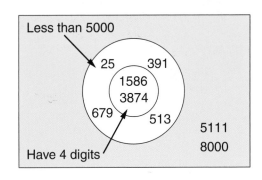

➤ Use a Venn diagram with overlapping circles.

The numbers in the left circle are greater than 2000.
The numbers in the right circle are odd.
The number in the overlap is greater than 2000 and is also odd.
1586 is not greater than 2000 and it is not odd, so it is outside the circles.

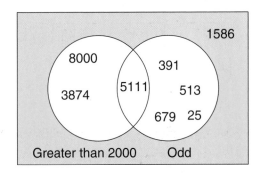

➤ Use a **Carroll diagram**.

In the first row, all the numbers are odd.
In the second row, all the numbers are not odd.
Numbers whose digits add to less than 10 are in the first column.
Numbers whose digits add to 10 or more are in the second column.

	Digits add to less than 10	Digits add to 10 or more
Odd	5111 513 25	391 679
Not odd	8000	1586 3874

8000 is not odd and its digits add to less than 10.

Practice

1. **a)** Sort these numbers in a Venn diagram.
Use the attributes: "Even" and "Greater than 500."
494, 627, 806, 213, 529, 740, 89, 2017

   b) Write one more number in each part of your Venn diagram.
Circle each number you write.

2. Copy the Venn diagram.

   a) How have these numbers been sorted?
Label each circle.

   b) Explain why each number
in the Venn diagram belongs
where it is placed.

   c) Write these numbers in the Venn
diagram: 920, 2563, 5808, 246

   d) What other numbers could you write in
each part of the Venn diagram?

		624 5168
567	603	20
845	3201	7026
931		804

3. Copy this Venn diagram.
Use the Venn diagram to sort
these numbers.
4725, 9902, 2477, 385,
7265, 6608, 2945, 776

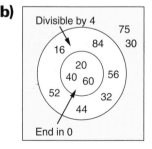

4. Describe what each Venn diagram shows.
How are the numbers related?
Explain the arrangement of the circles in each diagram.

a)

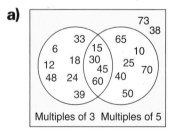

Multiples of 3 Multiples of 5

b)

Divisible by 4

75
16   84   30
20
40 60   56
52   32
44

End in 0

c)

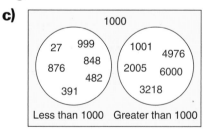

Less than 1000   Greater than 1000

5. Joe and Cher work at the dairy bar.
Joe worked on April 3rd
and every third day after that.
Cher worked on April 4th
and every fourth day after that.
Use a Venn diagram to find the dates in
April that Joe and Cher worked together.

6. Copy the Carroll diagram below.
Then sort these numbers in the diagram.
15, 50, 24, 30, 45, 19

	Even	Odd
Multiples of 3	6 36 12 42	9 21 27 39
Not multiples of 3	8 16 44 74	35 53 67 17

7. a) Copy this Carroll diagram.
Sort these numbers in
the Carroll diagram:
15, 36, 60, 99, 83, 55, 74, 85, 17, 42

	Is divisible by 5	Is not divisible by 5
Is even		
Is not even		

b) Write another number in each
box in the Carroll diagram.

c) Use the numbers from
parts a and b.
Sort the numbers in a Venn diagram.
Use the attributes "Even" and "Divisible by 5."

d) Do your Carroll diagram and your
Venn diagram show the same information?
Explain how you know.

Reflect

You have used Venn diagrams and Carroll diagrams.
How do you decide which diagram to use to sort a set of numbers?

Estimating Sums

Do you think doctors use an estimate when they prescribe medicine?

When you don't need an exact answer, you estimate. When would you use an estimate?

When you estimate a sum, you find a number that is close to the sum.

Explore

➤ About how much will it cost to buy a TV set and a DVD player?

➤ What could you buy if you had $700 to spend?

Estimate to find out. Record your answers.

ZAP electronics superstore

TV set	$589
DVD player	$204
VCR	$162
Computer	$998
Printer	$126
Keyboard	$119

SAVE $$$

Show *and* Share

Compare your answers with those of another pair of students.
Are your estimates higher or lower? Explain.
What strategies did you use to estimate?

Connect

An electronics store had 395 customers on Friday
and 452 customers on Saturday.
About how many customers did the store have for those 2 days?

When a question asks "about how many," you can estimate.

When you estimate, you use numbers that are close but easier to work with.

➤ Estimate: 395 + 452
You could write each number to the closest 100.
395 is closest to 400.
452 is closest to 500.

Add the numbers: 400 + 500 = 900
The store had about 900 customers for the 2 days.

Since 400 > 395 and 500 > 452, the estimate is high.

➤ Estimate: 395 + 452
You could use front-end estimation.
Add the first digits of the numbers.
395 + **4**52 is about 300 + 400 = 700.

For a closer estimate of 395 + 452:
Think about 95 + 52.
This is about 100 + 50 = 150.
Add 150 to the front-end estimate.
So, 395 + 452 is about 700 + 150 = 850.
The store had about 850 customers for the 2 days.

Practice

1. How many digits do you think each answer will have?
   Explain.
   a) 714 + 621   b) 1375 + 2496   c) 265 + 661

2. Raji estimated each sum.
   Is each estimate high or low?
   How do you know?
   a) 517 + 475 as 900   b) 4316 + 3442 as 7000
   c) 5678 + 1785 as 8000   d) 4056 + 359 as 4300

3. Estimate each sum.
   Explain your strategy.
   a) 71 + 847   b) 165 + 72   c) 5192 + 2192
   d) 189 + 2148   e) 982 + 828   f) 5307 + 88

4. Sam wants a lunch with less than 1000 Calories.
He has a hamburger with 445 Calories,
an apple pie with 405 Calories,
and ice cream with 270 Calories.
a) About how many Calories are in the lunch?
b) Did Sam make his goal? Explain.

5. Write a story problem where you would *not* use
estimation to solve it.
Explain why you would not estimate.

6. Look at these two addition questions:
4491 + 4491   4517 + 4517

a) Estimate each sum by writing each number to the closest 1000.
b) Use a calculator.
Are the two sums as different as the estimates make
them seem? Explain.
c) How might you get a better estimate for each sum?

7. When you estimate to add, how can you tell if the
estimated sum is greater than or less than the exact sum?

8. The estimated sum of two numbers is 600.
What might the numbers be?
Show your work.

9. Look at the list of numbers: 538, 476, 852, 938, 725
Which 2 numbers will give the sum that is closest to each
number below?
Show your work.
a) 1000   **b)** 1800

Reflect

Describe a situation in which you would estimate
a sum rather than find the exact answer.

Using Mental Math to Add

Explore

Students from two schools went on a field trip.
There were 227 students in one school,
and 134 students in the other school.
How many students went on the field trip?

Use mental math to find out.
Record your answer.

Show *and* Share

Share your strategies for adding with another pair of students.

Connect

There are many ways to use mental math to add.

➤ Winston uses the strategy of make a "friendly" number to add 198 + 343.

➤ Trang uses the strategy of counting on to add 170 + 348.

I know 198 + 2 is 200.
I take the 2 from 343.
That leaves 341.
200 + 341 is 541.
So, 198 + 343 is 541.

First I add 170
and 300. That makes 470.
Next I count on by 10 four
times: 470, 480, 490, 500, 510
Then I add 8: 510 + 8 = 518
So, 170 + 348 = 518

➤ Alexia uses the strategy of adding on from left to right to add 353 + 260.

> I'll start by
> adding the hundreds –
> 3 hundred ... 5 hundred
> Next I add the tens –
> 5 hundred fifty ...6 hundred ten
> Then the ones –
> 6 hundred thirteen
> So, 353 + 260 = 613

Practice

Use mental math.

1. Add. You choose the strategy.
   a) 179 + 234   b) 4266 + 4313   c) 4002 + 5336   d) 723 + 856
   e) 348 + 434   f) 536 + 299   g) 184 + 2301   h) 7620 + 95

2. Add.
   a) 263 + 328   b) 1439 + 2544   c) 190 + 943   d) 3998 + 432
   e) 691 + 180   f) 270 + 438   g) 3218 + 579   h) 2548 + 1573

3. There were 168 children in the park on Friday morning.
   There were 273 different children in the park on Friday afternoon.
   How many children were in the park on Friday?

4. Make up an addition problem you can solve using mental math.
   Describe the strategy you used to solve the problem.

Reflect

You know several strategies to add mentally.
Which is your favourite strategy?
Can you always use it?
Use words and numbers to explain.

Adding 3-Digit Numbers

Explore

Madhu uses the two sets of building blocks together.
How many pieces does she have?

Use any materials or strategies you like.
Use pictures, numbers, or words to
show your work.

Show *and* Share

Share your results with another pair of students.
Did you use the same strategies? Explain.
What other strategy could you use to solve the problem?

Connect

One jigsaw puzzle has 357 pieces.
Another puzzle has 275 pieces.
How many pieces are there altogether?

Add: 357 + 275
Here are different strategies students used.

➤ Abigail added from left to right.
   Add the hundreds (300 + 200).
   Add the tens (50 + 70).
   Add the ones (7 + 5).
   Add the sums.
   357 + 275 = 632

```
  357
+ 275
  500
  120
   12
  632
```

LESSON FOCUS | Use personal strategies to add 3-digit numbers.

51

➤ Sayid used Base Ten Blocks on a place-value mat to add 357 + 275.

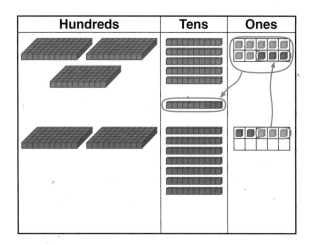

Sayid added the ones to get 12 ones.
He traded 10 ones for 1 ten.

$$\begin{array}{r} 1 \\ 357 \\ + 275 \\ \hline 2 \end{array}$$

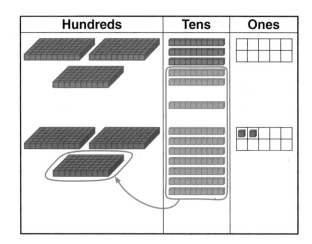

That made 13 tens 2 ones
Sayid traded 10 tens for 1 hundred.

$$\begin{array}{r} 11 \\ 357 \\ + 275 \\ \hline 32 \end{array}$$

That made 6 hundreds 3 tens

$$\begin{array}{r} 11 \\ 357 \\ + 275 \\ \hline 632 \end{array}$$

357 + 275 = 632

There are 632 jigsaw puzzle pieces.

1. Estimate first.
   Then add the numbers for which the sum will be greater than 600.
   a) 503 + 365   **b)** 817 + 179
   c) 199 + 52   **d)** 765 + 384

2. Estimate first.
   Then use any strategy you wish to find each sum.
   a) 384 + 765   **b)** 174 + 89
   c) 305 + 168   **d)** 491 + 256

3. A video store rented 165 more DVDs than video games.
   The store rented 258 video games.
   How many DVDs did the store rent?

4. The sum of two numbers is 756.
   What might the numbers be?
   How do you know?
   Can you find more than one pair of numbers?
   Explain.

5. What is the greatest number you can
   add to 365 *without* having to regroup
   in any place?
   Show your work.

6. Rahim visits golf courses to look for stray balls.
   He collected 209 golf balls last month.
   He collected 389 golf balls this week.
   How many golf balls did Rahim collect in all?

7. Janny has two sticker books.
   She has 488 stickers in one book and
   374 stickers in the other book.
   How many stickers does Janny have altogether?

8. Two hundred ninety-six students went skating on Monday.
Three hundred eight students went skating on Wednesday.
How many students went skating on the two days?

9. Carlotta delivered 427 flyers.
Chad delivered 583 flyers.
How many flyers did Carlotta and Chad deliver in all?

10. a) Write a story problem that can be solved by adding two
3-digit numbers.
 b) Write an equation for your story problem.
Solve the equation.
Show your work.

11. Each letter in this sum represents
a different digit.
What is the value of each letter?
How do you know?

```
  S E E
+ Y O U
-------
S O O N
```

Math Link

History

The abacus is used for counting.
You can add, subtract, multiply, and
divide with it.
The abacus was invented in China
over 800 years ago.
In North America, blind children are
taught to use the abacus.

Reflect

Which strategy do you prefer to add two 3-digit numbers?
Use an example to explain.

Adding 4-Digit Numbers

People set up dominoes in patterns.
So, when 1 domino topples, the rest topple.

Explore

There are 1275 dominoes in one set-up.
There are 2168 dominoes in another set-up.
How many dominoes are there altogether?

Use what you know about adding 3-digit numbers
to solve this problem. Show your work.

Show *and* Share

Share your strategy with another pair of students.
How could you add without using Base Ten Blocks?

Connect

Scott Suko is a world famous "domino toppler."
On his Website, there are photos of his set-ups.
One of Scott's set-ups had 1976 dominoes.
Another set-up had 2868 dominoes.
How many dominoes were there altogether?

Add: 1976 + 2868

Here are different strategies students use to solve the problem.

➤ Joel adds from right to left.

Six ones and 8 ones are 14 ones. Regroup 14 ones as 1 ten 4 ones.

One ten plus 7 tens plus 6 tens are 14 tens. Regroup 14 tens as 1 hundred 4 tens.

One hundred plus 9 hundreds plus 8 hundreds are 18 hundreds. Regroup 18 hundreds as 1 thousand 8 hundreds.

One thousand plus 1 thousand plus 2 thousands are 4 thousands.

So, 1976 + 2868 = 4844

➤ Zena uses column addition.

Add each column.

1000s	100s	10s	1s
1 +2	9 8	7 6	6 8
3	17	13	14
4	7	13	14
4	8	3	14
4	8	4	4

17 hundreds is 1 thousand 7 hundred.
Adjust the 1000s and 100s.

13 tens is 1 hundred 3 tens.
Adjust the 100s and 10s.

14 ones is 1 ten 4 ones.
Adjust the 10s and 1s.

1976 + 2868 = 4844

There are 4844 dominoes altogether.

The students estimate to check that the sum is reasonable.
1976 is close to 2000.
2868 is close to 3000.

2000 + 3000 = 5000
Since 4844 is close to 5000, the sum is reasonable.

Practice

1. Find each sum. Estimate to check.
   a) 4167 + 2534   b) 3974 + 4382   c) 5287 + 3756

2. Add. How do you know each sum is reasonable?
   a)   7865
      + 1987

   b)   3198
      + 6751

   c)   9999
      + 324

3. a) Write a story problem that could be solved by adding:
4267 + 1398

b) Estimate the sum. How did you get your estimate?

c) Is your estimate high or low? How do you know?

d) Find the sum. What strategy did you use?

e) How do you know your answer is reasonable?
Show your work.

4. Three thousand six hundred forty-two people went to the
Fall Fair on Friday.
Four thousand seven hundred ninety-five people went on Saturday.
How many people went to the Fall Fair on these 2 days?

5. The sum of two 4-digit numbers is 3456.
What might the two numbers be? Explain.

6. Jake guesses there are 2193 jellybeans in the jar.
Helena's guess is 1943 greater than Jake's guess.
What is Helena's guess?

7. Patsi's Girl Guide group collects pop can tabs.
The group collected 4594 tabs last year and
4406 tabs this year.
How many tabs did the group collect in the
two years?

8. The Wong family grows apples and pears in its orchards.
This fall, the Wongs picked 3265 baskets of apples and
2144 baskets of pears.
How many baskets of fruit did the Wongs pick?

9. Babu has saved 2363 pennies.
Serena has saved 3048 pennies.
How many pennies did Babu and Serena save altogether?

Reflect

How do you keep track of digits with the same place value
when you add? Use numbers and words to explain.

Estimating Differences

Explore

An arena has 594 seats.
Three hundred eight tickets have been sold for a concert.
About how many tickets are left?
Estimate to find out. Record your answer.

GREENBELT
LIVE IN CONCERT

Sat 7:00 at Midtown Arena

New CD

Tickets available at box office
$10, $20, and $30

Show *and* Share

Compare your estimate with that of another pair of students.
Did the strategies you used affect your answers? Explain.

Connect

➤ Estimate: 612 − 387
Write each number to the closest 100.
612 is closer to 600 than 700.
387 is closer to 400 than 300.
Subtract:
600 − 400 = 200
So, 612 − 387
is about 200.

You get a closer
estimate if you write only
one number to the closest 100.
Write 387 as 400.
612 − 400 = 212
So, 612 − 387 is about 212.

Make friendly
numbers before
you subtract.

➤ Estimate: 3274 − 1186
Use front-end estimation.
3274 → 3000
1186 → 1000
3000 − 1000 = 2000
So, 3274 − 1186 is about 2000.

Use the digits in the
thousands place.
Replace the other
digits with zeros.

➤ Estimate: 824 − 479
Write the number you subtract to the closest 100.
824 − 500 = 324
So, 824 − 479 is about 324.

500 is greater than 479, so this estimate is low.

Practice

1. Use any strategy you wish to estimate each difference.
   a) 871 − 263   **b)** 610 − 429
   c) 734 − 591   **d)** 9907 − 6254

2. Kyle estimated each difference.
   Is each estimate high or low?
   How do you know?
   a) 576 − 392 as 100   **b)** 911 − 188 as 800   **c)** 7361 − 1872 as 6000

3. Estimate each difference by writing each number to the closest 100.
   a) 983 − 407   **b)** 7720 − 6953
   c) 918 − 75   **d)** 447 − 293

4. Charlotte looks at this survey.
   She says, "About 300 more students chose biking over walking."
   a) How might Charlotte have estimated? Explain.
   b) Is the estimate high or low? Explain.
   c) What might have been a better way to estimate?

Favourite Exercise

| Biking | 378 students |
| Walking | 124 students |

5. Write a subtraction problem that can be solved by estimating.
   Solve the problem.
   Show your work.

6. The estimated difference of two numbers is 300.
What might the numbers be?
Explain how you found the numbers.

7. Describe a situation where you would estimate
rather than find the exact answer to a
subtraction problem.

8. Estimate each difference by using front-end estimation.
   a) 763 − 419   **b)** 7647 − 2991
   c) 988 − 462   **d)** 9411 − 6231

9. Hans had 528 paper clips.
He gave 257 of them to Gertie.
About how many paper clips does Hans have left?

10. Nikki's school has 491 students.
David's school has 703 students.
About how many more students
does David's school have
than Nikki's school?

11. The telephone was invented in 1876.
About how many years ago was that?

12. The CN Tower is 553 m tall.
The Empire State Building is 380 m tall.
About how much taller is the CN Tower than
the Empire State Building?

Reflect

When does writing the numbers to the closest 100 not
give a good estimate when you subtract?
Use words and numbers to explain.

9 Using Mental Math to Subtract

Explore

Anita used 354 cards to make a house of cards.
Christopher used 198 to make his house.
How many more cards are in Anita's house
than Christopher's?
Use mental math to find out. Record your answer.

Show and Share

Share the strategy you used with another pair of students.

Connect

➤ Use mental math to subtract: 516 − 299
Bruce uses the strategy of make a friendly number.
He adds 1 to 299 to make 300.
He adds 1 to 516 to make 517.
He thinks: 517 − 300 = 217
So, 516 − 299 = 217

If I add 1 to each number, the answer will not change.

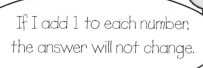

299 ————————→ 516
300   400   500 517

➤ Use mental math to subtract: 347 − 195
Marly uses a friendly number.
She subtracts 200 instead of 195.
She thinks: 347 − 200 = 147
Then she adds 5.
147 + 5 = 152
So, 347 − 195 = 152

I took away 200 instead of 195. Since I took away 5 too many, I added the 5 at the end.

➤ Use mental math to subtract: 432 − 220
Harry uses the strategy of "counting on."
He counts on from 220.

Count: 220,   320,   420,   430,   432

       + 100   + 100   + 10   + 2   = 212

So, 432 − 220 = 212

I use this strategy when there are not too many steps to count on.

Practice

Use mental math.

1. Subtract. Which strategy did you use each time?
   a) 536 − 399   **b)** 6352 − 1887   **c)** 822 − 216   **d)** 4231 − 2984

2. Subtract 715 − 197 mentally as many different ways as you can.
   Which strategy was easiest? Explain.

3. How much change will you get from $1000 when you
   buy something that costs $680?
   How do you know?

4. The answer to a subtraction problem is 127.
   Use mental math to find what the problem might be.
   Write as many different problems as you can.
   Show your work.

5. Write a subtraction problem you can solve
   using mental math.
   Solve the problem.

Reflect

Which mental math strategy is easiest for you?
Use words and numbers to explain.

Subtracting 3-Digit Numbers

Explore

There are 430 students at Hirondelle School.
Two hundred sixty-five students are boys.
How many students are girls?

Use any materials or strategies you like.
Use pictures, numbers, or words to
show your work.

Show *and* Share

Share your strategy with another pair of students.
What other strategy could you use to solve the problem?

Connect

Here are different strategies students used to subtract.

➤ Frankie makes a friendly number, to subtract: 565 − 317

$$
\begin{array}{ccccccc}
565 & -7 & 558 & -10 & 548 \\
-317 & -7 & -310 & -10 & -300 \\
\hline
& & & & 248
\end{array}
$$

I took away 7
to make 310.
Then I took away 10
to make 300.

So, 565 − 317 = 248

➤ Kada uses Base Ten Blocks
on a place-value mat
to subtract: 400 − 286

You cannot take 6 ones
from 0 ones.
There are no tens to trade.
So, I trade 1 hundred for
10 tens.

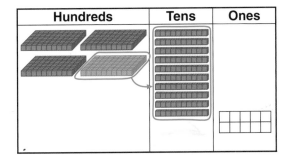

I trade 1 ten
for 10 ones.

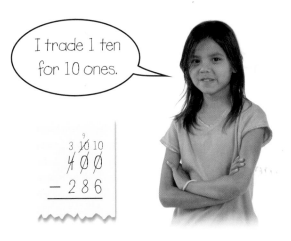

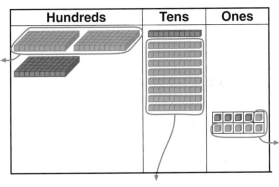

I take away 6 ones,
8 tens, and
2 hundreds.

So, 400 − 286 = 114

Kada checks by adding.

Add: 286 + 114
The sum should be 400.
Since 286 + 114 is 400, the answer is correct.

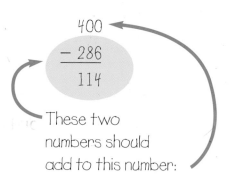

These two
numbers should
add to this number:

➤ Jan uses counting on to subtract: 622 − 397
She counts on from 397 to 622.

| +3 | +100 | +100 | +22 |

Count: 397   400   500   600   622

Think: 3   ... 103   ... 203   ... 225

So, 622 − 397 = 225

As I count on,
I add in my head.

Practice

1. Subtract.
   What patterns do you see in the questions and answers?
   a) 857 − 100   **b)** 857 − 200   **c)** 857 − 300   **d)** 857 − 400

2. Estimate first. Then subtract the numbers for which the answer will be less than 200.

   a)   255
        − 76

   b)   426
        − 158

   c)   678
        − 298

   d)   382
        − 192

3. Subtract. How do you know each answer is reasonable?

   a) 565 − 317   b) 700 − 189   c) 101 − 96   d) 861 − 178

4. Sadiq read 315 pages. Laura read 248 pages. How many more pages does Laura need to read to catch up with Sadiq?

5. The largest gorilla has a mass of about 275 kg. The largest orangutan has a mass of about 90 kg. What is the difference in their masses?

6. The world records for barrel jumps are held by Canadians. The longest barrel jump by a woman is 670 cm. The longest barrel jump by a man is 882 cm. How much farther is the man's jump? How do you know your answer is reasonable? Show your work.

7. a) The answer to a subtraction problem is 375. What might the problem be? Write as many problems as you can.

   b) The answer to an addition problem is 375. What might the problem be? Write as many problems as you can.

Reflect

Explain why you can check a subtraction problem by adding.

Strategies Toolkit

Explore

Fiona is 5 cm taller than Zac.
Together their heights total 299 cm.
How tall is Fiona? How tall is Zac?

Work together to solve this problem.
Use any materials you think will help.

Show *and* Share

Describe the strategy you used to
solve the problem.

Connect

Yael and Victor collect postcards.
Yael has 10 more postcards than Victor.
Together, they have 420 postcards.
How many postcards does each person have?

What do you know?
- There are 420 postcards in all.
- Yael has 10 more postcards
  than Victor.

Think of a strategy to help you
solve the problem.
- You can **make an organized list**.
- Find two numbers that add to 420.
  One number must be 10 more than
  the other.

Strategies

- Make a table.
- Use a model.
- Draw a diagram.
- Solve a simpler
  problem.
- Work backward.
- Guess and test.
- Make an organized
  list.
- Use a pattern.

LESSON FOCUS | Interpret a problem and select an appropriate strategy.

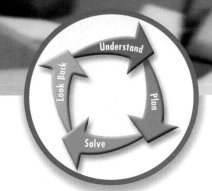

Make an organized list to show the numbers.

Choose a number for Yael's postcards; such as 220.

Subtract 220 from the total to find Victor's postcards:

420 − 220 is 200 postcards for Victor.

Subtract the numbers of postcards:

220 − 200 = 20 This is too high.

Try 1 less for Yael and 1 more for Victor.

Yael's postcards	Victor's postcards	Difference	
220	200	220 − 200 = 20	Too high
219	201	219 − 201 = 18	Too high

Continue this strategy until the difference is 10.

Could you have tried 2 less for Yael and 2 more for Victor instead? Explain.

Practice

Choose one of the
Strategies

1. The Huda family picked 800 cucumbers in two days.
   They picked 124 more cucumbers on the first day than on the second day.
   How many cucumbers did the family pick each day?

2. Raphie has 90 cents in dimes and nickels.
   She has the same number of each coin.
   How many of each coin does Raphie have?

Reflect

What is the difference between "making a list" and "making an organized list"?

Which is the better strategy for solving problems? Explain.

Subtracting 4-Digit Numbers

Explore

Matthew's school created a Website.
One day, the site had 1531 visitors. The next day it had 867 visitors.
How many more people visited the site the first day?

Use what you know about subtracting
3-digit numbers to solve this problem.

Show *and* Share

Share your solution with another pair of students.
How did you subtract without using Base Ten Blocks?

Connect

How many more people visited the Website
on Friday than on Saturday?

Subtract: 2031 − 856

Day	Visitors to Website
Friday	2031
Saturday	856

Here are the strategies some students used to solve the problem.

➤ Rod explains how he subtracts from right to left.

"You cannot take 6 ones
from 1 one.
Regroup 1 ten as
10 ones.
Then, subtract the ones."

$$\begin{array}{r} {}^{2}\,{}^{11}\\ 20\cancel{3}\cancel{1}\\ -\ 856\\ \hline 5 \end{array}$$

"You cannot take 5 tens from 2 tens.
There are no hundreds to regroup.
So, regroup 1 thousand as
10 hundreds.
Then, regroup 1 hundred as 10 tens."

$$\begin{array}{r} {}^{9}\ {}^{12}\\ {}^{1}\,{}^{10}\,{}^{2}\,{}^{11}\\ \cancel{2}\cancel{0}\cancel{3}\cancel{1}\\ -\ 856\\ \hline 5 \end{array}$$

"Then, subtract the tens.
Subtract the hundreds.
Subtract the thousands."

$$\begin{array}{r} {}^{9}\ {}^{12}\\ {}^{1}\,{}^{10}\,{}^{2}\,{}^{11}\\ \cancel{2}\cancel{0}\cancel{3}\cancel{1}\\ -\ 856\\ \hline 1175 \end{array}$$

There were 1175 more visitors on Friday than on Saturday.

➤ Kate takes 31 away from each number then adds, to make friendly numbers.

2031 − 856

2031	−31	2000	+75	2075
− 856	−31	−825	+75	−900
				1175

I know that if I take 31 away from each number, it does not change the answer.

Kate checks her answer by estimating.
2031 is closest to 2000.
856 is closest to 900.
2000 − 900 = 1100

1100 is close to 1175; so, the answer is reasonable.

➤ Maksim uses the strategy of make a friendly number, then compensates.

2031 − 856

2031 − 856 → 2031 − 900 = 1131
900 − 856 = 44
1131 + 44 = 1175
So, 2031 − 856 = 1175

I subtracted 900 instead of 856. That's 44 too many. So I added 44 at the end.

Practice

1. Estimate, then subtract.
   Is each answer reasonable? Explain.
   a) 8274 − 3596 **b)** 6328 − 937 **c)** 4028 − 1639

2. Subtract. Check your answer.
   a) 3102   **b)** 5287   **c)** 7000
        − 1428   − 931   − 2476

3. Subtract.
   a) 7130 − 2864   **b)** 9345 − 6898   **c)** 6005 − 4816

4. Subtract.
   a) Seven thousand one minus three hundred fifty-six
   b) Eight thousand twelve minus four thousand two hundred twenty-eight

5. Is it possible to subtract a 3-digit number from a
   4-digit number and get a 4-digit number as the answer?
   A 3-digit number as the answer?
   A 2-digit number as the answer?
   A 1-digit number as the answer?
   Give an example for each possible answer.
   Show your work.

6. In 1215, the Magna Carta was signed.
   How many years ago was that?

7. Use eight different digits from 1 to 9.

   ☐☐☐☐
   − ☐☐☐☐

   a) What is the greatest difference you can make?
   b) What is the least difference you can make?
   c) How do you know the answer you found in
        part a is the greatest? In part b is the least?

8. Each letter in this problem
   represents a different digit
   from 0 to 9.
   What is the value of each letter?
   How do you know?

   S H H H
   −   S
   ───────
     Z Z Z

Reflect

How do you keep track of numbers with the same place value
when you subtract? Use numbers and words to explain.

Solving Addition and Subtraction Problems

Some athletes take part in stair-climbing races.
Some people climb the stairs to raise money for charity.

CN Tower
1776 Steps

Central Park
Tower, Australia
1236 Steps

Menara Tower,
Malaysia
2058 Steps

Mei participated in stair-climbing events at the Menara Tower,
the Central Park Tower, and the CN Tower.

➤ How many steps did she climb altogether?

➤ Make up another problem, then solve it.

Show *and* Share

Trade problems with another pair of students. Solve their problem.
Compare your strategies for solving the problems.

Connect

Ahmal started a business selling computer parts.
He opened a bank account with $1776.
In his first two weeks he deposited $1236 and
$2109 into the account.

LESSON FOCUS | Use different strategies to add and subtract more than 2 numbers.

➤ How much did he have in the account altogether?
Add: 1776 + 1236 + 2109

	1776
Add from left to right.	1236
	+ 2109
Add the thousands (1000 + 1000 + 2000).	4000
Add the hundreds (700 + 200 + 100).	1000
Add the tens (70 + 30 + 0).	100
Add the ones (6 + 6 + 9).	21
Add the sums.	5121

Ahmal had $5121 altogether in the account.

➤ In the third week, Ahmal had to pay 2 bills
of $1041 and $650.
How much did Ahmal have in his account
after paying the bills?

Ahmal started with $5121 in his account.

- Subtract: 5121 − 1041
  Subtract from right to left.

$$
\begin{array}{r}
^{0\ 12}\!5\cancel{1}\cancel{2}1 \\
-\ 1041 \\
\hline
4080
\end{array}
$$

- Then subtract 650 from the result.

$$
\begin{array}{r}
^{3\ 10}\!4\cancel{0}80 \\
-\ 650 \\
\hline
3430
\end{array}
$$

Ahmal had $3430 in his account after paying his bills.

Practice

1. Find each sum.
   a) 1175 + 3241 + 829   b) 2456 + 3727 + 1104
   c) 4782 + 543 + 2368   d) 3040 + 4307 + 5198

2. Juan drives a truck. On Monday,
he left Prince George to drive 1639 km
to Whitehorse. On Wednesday,
he left to drive 1222 km to Inuvik.
On Saturday, he left to drive 3149 km to Yellowknife.
How far did Juan travel altogether?

3. The Lees drove 1431 km to their summer home.
On their return, they took the same route.
They drove 613 km the first day and 486 km on the second day.
How far would the Lees have to drive on the third day
to get home?

4. Kay spilled a drink on her homework.
Copy and complete the addition.
Find the digits that are covered.
Explain how you know.

```
   3 6 □ 5
   △ 9 7 4
 + 1 1 5 ○
 ─────────
   7 * 1 5
```

5. The sum of 3 numbers is 8196.
One of the numbers is 988.
What might the other two numbers be?
How do you know?

6. At the beginning of the month,
Anne had $2340 in her bank account.
Anne deposited $936 one day and $94 another day.
Anne took out $790 the next week.
How much did Anne have in her account then?

7. Find a number you can add to 6274 so you have to
regroup ones, tens, and hundreds.
Can you find more than one number? Explain.

Reflect

When you add three numbers, does the order in which
you add the numbers matter?
Does the same rule apply to subtraction? Explain.

Show What You Know

1

1. The highest score in a Scrabble game is 1049.
Write this number in words and in expanded form.

2. Explain the meaning of each digit in the number 8888.

3. Write each number in standard form, then in a place-value chart.
   a) eight thousand twenty-six   **b)** 6000 + 800 + 7

4. Draw a picture to represent each number in question 3.

2

5. Write these numbers in order from least to greatest.
5242, 5232, 5223

3

6. Use a Venn diagram to sort these numbers:
3057, 555, 2454, 333, 636, 22, 4444
You choose the attributes.

4
8

7. Estimate each sum or difference.
   a) 680 + 213   **b)** 2761 − 1780   **c)** 176 + 412
   d) 597 − 237   **e)** 1276 + 2566   **f)** 911 − 499

5
9

8. Use mental math to add or subtract.
   a) 2567 + 1724   **b)** 385 − 189   **c)** 247 + 338   **d)** 4210 − 2983

4
8

9. For a Read-A-Thon, Natalie read 786 pages.
Kevin read 815 pages. Mario read 623 pages.
Altogether, they read over 2000 pages.
   a) Is 2000 exact or an estimate? How do you know?
   b) About how many more pages did Kevin read than Mario?

6
7
10
12

10. Add or subtract. How do you know your answers are reasonable?

a) 2211	b) 452	c) 800	d) 4579
− 878	+ 348	− 298	+ 3975

e) 762	f) 737	g) 993	h) 9843
− 304	+ 843	+ 5002	− 4213

11. The Musicians, a rock group, had 3 concerts last month. The first concert had an audience of 4356. The second concert had an audience of 3295. The third concert had an audience of 2964. How many people attended the concerts altogether?

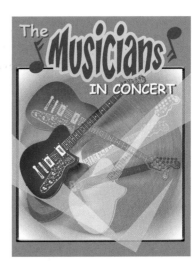

12. Refer to question 11. The first Musicians concert was a promotion night.
Seven hundred forty-six tickets were given away through radio contests.
Three hundred twelve tickets were given away through Internet promotions.
The rest of the tickets were purchased by fans.
How many tickets were purchased for the first concert?

Use the following information to answer questions 13 to 15.
Container A holds 2500 unit cubes.
Container B holds 1875 unit cubes.
Both containers are full.

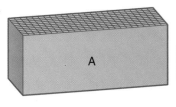

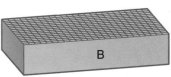

8

13. How many cubes do the two containers hold altogether?

13

14. Rhonda takes 725 cubes from container A. Then Marilyn takes 925 cubes, and Everett takes 375 cubes. How many cubes are left in container A?

15. Is there enough room now in container A to hold the cubes from container B? Explain.

UNIT
2 Learning Goals

☑ recognize and read numbers from 1 to 10 000
☑ read and write numbers in standard form, expanded form, and written form
☑ compare and order numbers
☑ use diagrams to show relationships
☑ estimate sums and differences
☑ add and subtract 3-digit and 4-digit numbers mentally
☑ use personal strategies to add and subtract
☑ pose and solve problems

Those Amazing Elephants

Kamala's Art

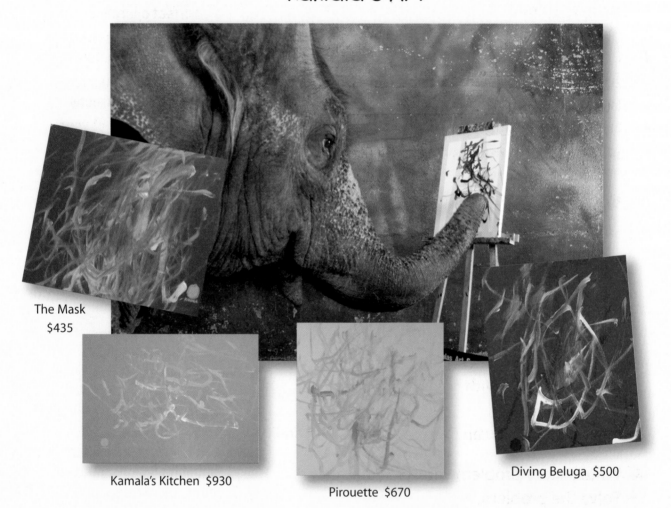

The Mask
$435

Kamala's Kitchen $930

Pirouette $670

Diving Beluga $500

1. The Calgary Zoo is home to 4 Asian elephants.
   Use the data in the table.
   a) Calculate the age of each elephant.
   b) Order the elephants from youngest
      to oldest.
   c) Kamala is Maharani's mother.
      How old was Kamala when
      Maharani was born?
   d) In what year will Spike be 50 years old?

Elephants' Year of Birth	
Name	**Year**
Swarna	1975
Maharani	1990
Spike	1981
Kamala	1975

78

Your work should show
- [✓] that you can choose the correct operation
- [✓] your thinking in words, numbers, or pictures
- [✓] how you added and subtracted correctly
- [✓] a clear record of your answers

2. Kamala has been named Canada's most famous animal.

She paints pictures that are sold for hundreds of dollars. The money earned from the sales of the paintings will be used to build a new home for the elephants.

Use the pictures on page 78.

a) Order the paintings from least to most expensive.

b) Choose 3 of Kamala's paintings you would like to own.
How much would you pay for them?

c) Find the difference in cost of the most expensive and least expensive paintings.

3. Elephants can pick up and drag very heavy objects. Oscar, an adult Asian elephant, can lift a 435-kg log with his tusks. He can drag a load of 1500 kg.
How much more can Oscar drag than he can lift?

4. Write a story problem about elephants.
Solve the problem.
Show your work.

Reflect on Your Learning

Write about the different strategies you know for adding and subtracting.

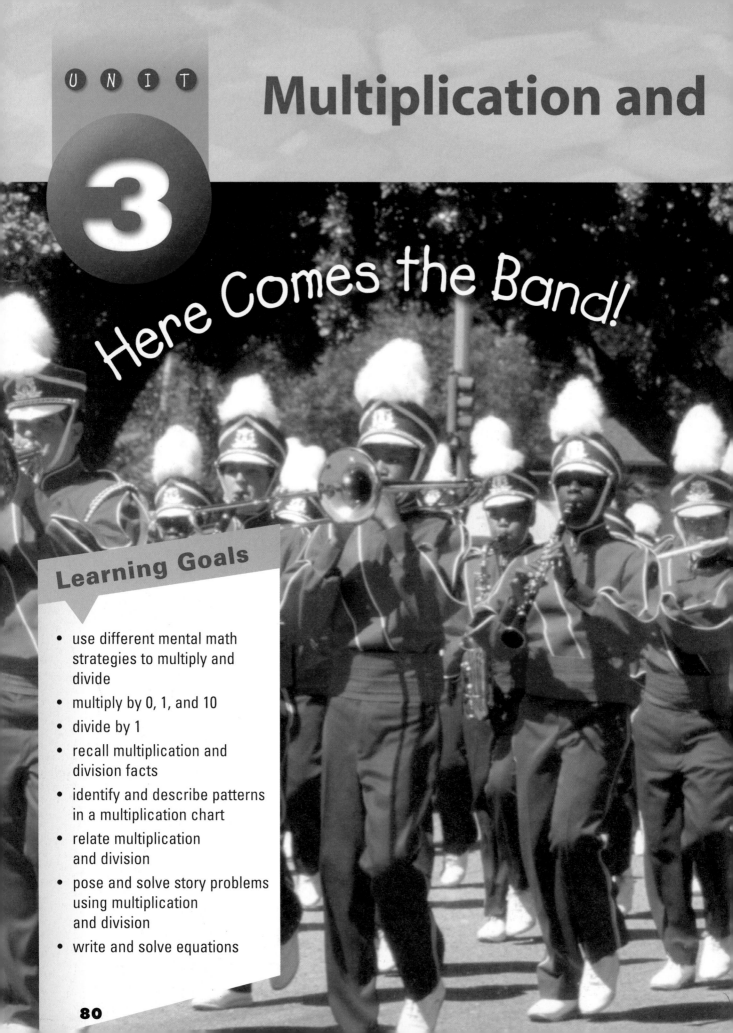

UNIT 3

Multiplication and

Here Comes the Band!

Learning Goals

- use different mental math strategies to multiply and divide
- multiply by 0, 1, and 10
- divide by 1
- recall multiplication and division facts
- identify and describe patterns in a multiplication chart
- relate multiplication and division
- pose and solve story problems using multiplication and division
- write and solve equations

Division Facts

Key Words

array

factor

product

row

column

multiples

related facts

In a parade, there are many bands.

- How many people are in this band?
- How did you find out?
- How else can you find out how many people are in the band?
- How many different ways can you find out how many are in the band?

Using Doubles to Multiply

An **array** shows objects arranged in equal rows.
You can use an array to multiply.
To find 2×3, make an array of 2 rows of 3.

2 rows of 3 counters
$2 \times 3 = 6$
↑
This is a multiplication fact.

Explore •

You will need 1-cm grid paper and scissors.

➤ Use the grid paper.
Draw an array for 3×4.
Cut out the array.
Record a multiplication fact for the array.

➤ Use your array and your partner's array to make a larger array.
Record a multiplication fact for this new array.
Arrange your arrays another way.
Record another multiplication fact.

➤ With your partner, choose one of these:
2×4 3×5 4×6
Cut out an array to show the fact.
Record a multiplication fact for the array.
Put your array together with your partner's array.
Record a multiplication fact for this new array.
Arrange the arrays another way.
Record a multiplication fact for that array.

Show *and* Share

Show your arrays and multiplication facts to another pair of students.
What do you notice when you double the size of an array?

Connect . . . .

In a multiplication fact, you multiply **factors** to get a **product**.

4 × 5 = 20

factor   factor   product

> When you double a number, you are multiplying by 2.

Doubling is a strategy you can use to multiply.
Here are three ways you can use doubling to multiply.

➤ Use doubling to multiply by 4.
To find 4 × 9:
First find 2 × 9, then double.

$2 \times 9 = 18$

$2 \times 9 = 18$

$18 + 18 = 36$

$4 \times 9 = 36$

➤ Use repeated doubling to multiply by 8.
To find 8 × 8:
First think of 2 × 8 = 16, then double, and then double again.

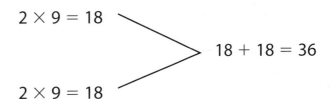

2×8 is $8 + 8 = 16$
So, $2 \times 8 = 16$

4×8 is double 2×8
$16 + 16 = 32$
So, $4 \times 8 = 32$

8×8 is double 4×8
$32 + 32 = 64$
So, $8 \times 8 = 64$

➤ Begin with a fact you know.
Double one of the factors, then multiply.

You know 2 × 3 = 6.
You can double the factor 2 to get 4.
4 × 3 = 12
Or, you can double the factor 3 to get 6.
2 × 6 = 12

When you double a factor, the product doubles.

Practice

Use counters when they help.

1. **a)** Write a multiplication fact for this array.
   b) Double one factor in the multiplication fact.
      Make an array for this new fact.
      Write a multiplication fact.

2. Six students can travel in 1 boat.
   a) How many students can travel in 2 boats?
   b) How many students can travel in 4 boats?
   c) How many students can travel in 8 boats?
   How do you know?

3. Multiply. What strategies did you use?
   a) 2 × 6   **b)** 4 × 6   **c)** 8 × 6
   d) 8 × 2   **e)** 8 × 4   **f)** 8 × 8

4. Write two facts that help you find the product of 6 × 8.

5. Find the product. What strategies did you use?
   a) 2 × 7   **b)** 4 × 7   **c)** 8 × 7
   d) 2 × 5   **e)** 4 × 5   **f)** 8 × 5

Measurement

To find how many
cents are in 8 nickels,
multiply: 8 × 5

6. How can you use 3 × 6 to find 6 × 6?
Show your work.

7. Find the product. What strategies did you use?
a) 7 × 4   **b)** 6 × 4   **c)** 7 × 6   **d)** 3 × 4

8. There are bicycles and wagons in the playground.
Each bicycle has 2 wheels.
Each wagon has 4 wheels.
Eli counted 28 wheels.
How many bicycles and wagons might there be?
How many different ways can you find the answer?
Use multiplication facts to show your work.

9. Write a multiplication story problem about wheels.
Solve your problem. Show your work.

10. What might each missing number be?
How many answers can you find?
a) □ × △ = 16   **b)** ○ × □ = 24

Reflect

How do you know that doubling the product of 2 × 6
is the same as finding 4 × 6?
Use words, numbers, or pictures to explain.

Multiplying by 1, by 0, and by 10

Explore ·

You will need paper plates and counters.

➤ Mark makes waffles for his family.
   He puts each waffle on a different plate.
   Mark uses 5 plates.
   How many waffles does Mark make?
   How can you use a multiplication
   fact to show this?

➤ Mark has 3 empty plates.
   How many waffles are on
   these plates?
   How can you use a multiplication
   fact to show this?

➤ Mark has 6 trays.
   He stacks 10 waffles on each tray.
   How many waffles does Mark stack?
   How can you use a multiplication
   fact to show this?

Show *and* Share

Show your work to another pair of classmates.
What is special about multiplying by 1?
What is special about multiplying by 0?
What is special about multiplying by 10?

➤ Marie made French crêpes for her friends.

She put 1 crêpe on each plate.
Marie used 6 plates.
How many crêpes did Marie make?

 6 groups of 1 is 6 × 1.

| 6 | × | 1 | = | 6 |
| plates | | crêpe | | crêpes in all |

Also, 1 × 6 = 6

When 1 is a factor, the product is always the other factor.

➤ Marie had 5 empty plates.

How many crêpes are on these plates?

 5 groups of 0 is 5 × 0.

| 5 | × | 0 | = | 0 |
| plates | | crêpes | | crêpes in all |

Also, 0 × 5 = 0

When 0 is a factor, the product is always 0.

➤ Marie made many crêpes and served them on trays.

She put 10 crêpes on each tray.
Marie used 3 trays.
How many crêpes did Marie serve?

Think: 3 groups of 10 is 3 × 10.

| 3 | × | 10 | = | 30 |
| trays | | crêpes | | crêpes in all |

Also, 10 × 3 = 30

Practice

1. Multiply.
   a) 0 × 6   **b)** 1 × 5   **c)** 0 × 7
   d) 1 × 3   **e)** 10 × 8   **f)** 10 × 3

2. Jessica buys 6 single-scoop ice cream cones.
   How many scoops of ice cream does Jessica buy?
   Show your answer using pictures and numbers.

3. Mario has 3 empty ice cream cones.
   How many scoops of ice cream does Mario have?
   Write a multiplication fact.

4. Solve each equation.
   a) 6 × □ = 6   **b)** □ × 3 = 0
   c) □ × 5 = 5   **d)** 2 × □ = 0
   e) □ × 10 = 90   **f)** 10 × □ = 70

5. Write a story problem that can be solved by multiplying by 0.
   Write an equation for the problem.
   Give your problem to a classmate to solve.

6. Mark puts these fruits on his waffle:
  - one strawberry
  - twice as many raspberries as strawberries
  - three times as many blueberries as raspberries

How many raspberries are on the waffle?
How many blueberries?
How do you know?

7. a) Is it easier to solve 24 × 1 or
  24 × 2? Explain.

b) Is it easier to solve 6 × 0 or
  66 × 0? Explain.

8. You have 6 dimes.
Your brother has 9 nickels.
Who has more money?
How do you know?
Show your work.

9. Which is greater:
a) 1 × 4 or 0 × 9?   **b)** 10 × 2 or 9 × 1?   **c)** 0 × 8 or 3 × 1?
How do you know?

10. a) Explain why the answer is always 0 when you multiply by 0.
b) Explain why the answer is always the other number when
  you multiply by 1.

Reflect

What can you say about:
- multiplying by 0?
- multiplying by 1?
- multiplying by 10?

Use words, pictures, or numbers to explain.

LESSON

3 Using Skip Counting to Multiply

You will need copies of these charts.

Hundred Chart

1	2	3	4	5	6	7	8	9	10
11	12	13	14	15	16	17	18	19	20
21	22	23	24	25	26	27	28	29	30
31	32	33	34	35	36	37	38	39	40
41	42	43	44	45	46	47	48	49	50
51	52	53	54	55	56	57	58	59	60
61	62	63	64	65	66	67	68	69	70
71	72	73	74	75	76	77	78	79	80
81	82	83	84	85	86	87	88	89	90
91	92	93	94	95	96	97	98	99	100

Multiplication Chart

x	1	2	3	4	5	6	7	8	9
1	1	2	3	4	5	6	7	8	9
2	2	4	6	8	10	12	14	16	18
3	3	6							
4	4	8							
5	5	10							
6	6	12							
7	7	14							
8	8	16							
9	9	18							

➤ Use the hundred chart.
  Start at 3 and count on by 3s.
  Use the numbers you count to fill in the next row
  and column in the multiplication chart.

➤ Repeat the activity.
  Start at 4 and count on by 4s.
  Start at 5 and count on by 5s.
  Count on by other numbers until you have filled in the chart.

What patterns do you see in the multiplication chart?

Show and Share

Share the patterns you found with another pair of students.
What patterns did you find in the ones digits? The tens digits?

Connect

Here are some ways to use skip counting to multiply mentally.

➤ Use skip counting patterns.
For example,
To find 7×6, skip count by 6 seven times:
6, 12, 18, 24, 30, 36, 42

*These numbers are **multiples** of 6.*

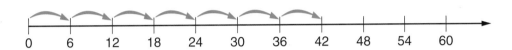

7 steps of 6 is 42.
$7 \times 6 = 42$

Another way is to skip count by 7 six times:
7, 14, 21, 28, 35, 42

These numbers are multiples of 7.

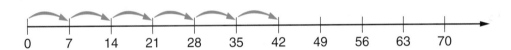

6 steps of 7 is 42.
$7 \times 6 = 42$

➤ Skip count from a known fact.
You know $9 \times 5 = 45$.
To find 9×6:
Skip count by 9 to add one more group of 9.

"45 ... then 54"

So, $9 \times 6 = 54$

To find 9×7:
You know $9 \times 5 = 45$.
Skip count by 9 to add two more groups of 9.

"45 ... then 54 ... then 63"

So, $9 \times 7 = 63$

1. Use a hundred chart
   a) Start at 4. Skip count by 4 to find the multiples of 4.
      List them.
   b) Start at 8. Skip count by 8 to find the multiples of 8.
      List them.
   c) Compare the numbers in both lists.
      What patterns do you see?

2. Skip count to find the missing numbers.
   a) 3, 6, 9, __, __, __, __
   b) 5, 10, 15, __, __, __, __
   c) 6, 12, 18, __, __, __, __
   d) 7, 14, 21, __, __, __, __

3. Use skip counting.
   Copy and complete these rows from a multiplication chart.

x	3	4	5	6	7	8	9
3	9	12					
4	12	16					

   Look at the completed rows for × 3 and × 4. How are they related?

4. Multiply.
   How does knowing the first fact help you complete the second fact?
   a) $6 \times 5 = 30$, then $7 \times 5 = \square$
   b) $3 \times 6 = 18$, then $4 \times 6 = \square$
   c) $5 \times 8 = 40$, then $6 \times 8 = \square$

5. There are 8 pairs of students ready to learn a French dance.
   How many students are there?
   Show your work.

6. Nicole collects nickels.
   Could she have 45 cents?
   Show your work.

7. Jillian knows that $3 \times 7 = 21$.
   How can she use that fact to find 5×7?

8. Play this game with a partner.

You will need a number cube labelled 4, 5, 6, 7, 8, 9; a counter; and a hundred chart.

Choose a target number between 50 and 100.

Put the counter on that number.

Roll the number cube to find the start number.

Use the same number to count on.

Have your partner say if you will "hit" the target when you count on.

Count on to check.

Or, explain how you know.

Trade roles.

Take turns to roll and check.

Game

9. Here is part of a multiplication chart.

Each shape in the chart represents a number.
  a) Which number does each shape represent?
  b) What strategies did you use to find each number?
  c) How are the rows for × 5 and × 6 related?
  Show your work.

×	5	6	7	8	9
5	25	30	★	40	45
6	30	☐	42	48	◯
7	35	42	49	△	63
8	♥	48	56	64	72
9	45	54	◯	72	☐

10. Use a calculator to count on.

Press the keys.

Record what you see on the screen.

Write about the patterns you see.
  a) Press: ⟨ON/C⟩ ⟨2⟩ ⟨+⟩ ⟨=⟩ ⟨=⟩ ⟨=⟩ ⟨=⟩ ⟨=⟩ ⟨=⟩ ⟨=⟩ ⟨=⟩ ⟨=⟩
  b) Press: ⟨ON/C⟩ ⟨5⟩ ⟨+⟩ ⟨=⟩ ⟨=⟩ ⟨=⟩ ⟨=⟩ ⟨=⟩ ⟨=⟩ ⟨=⟩ ⟨=⟩ ⟨=⟩
  c) Press: ⟨ON/C⟩ ⟨9⟩ ⟨+⟩ ⟨=⟩ ⟨=⟩ ⟨=⟩ ⟨=⟩ ⟨=⟩ ⟨=⟩ ⟨=⟩ ⟨=⟩ ⟨=⟩

Reflect

How does skip counting help you multiply?
Use words and numbers to explain.

4

Other Strategies for Multiplying

 **Explore** ·  Game

Play this game with a partner.

Cross-Out Product

You will need a number cube labelled 4, 5, 6, 7, 8, 9; and 2 copies of this game board.

4	6	9	12	14
18	20	24	27	32
35	40	42	45	49
56	63	64	72	81

➤ One person rolls the number cube to get a factor.
   Each person thinks of a multiplication fact that uses that factor.
   Record the fact you use.
   Cross out the product on your game board.

➤ Take turns to roll the number cube.

➤ The first person to cross out all the products is the winner.

Show *and* Share

Talk with your partner about the strategies you used to multiply.
What mental math strategies did you use?

Another way to multiply is to add groups to facts you know.

➤ Use facts with 2 to multiply by 3.
First, multiply by 2.
Then, add another group.
To find 3×7:

Think: $2 \times 7 = 14$

$14 + 7 = 21$

$1 \times 7 = 7$

So, $3 \times 7 = 21$

➤ Use facts with 5 to multiply by 6.
First, multiply by 5.
Then, add another group.
To find 6×7:

Think: $5 \times 7 = 35$

$35 + 7 = 42$

$1 \times 7 = 7$

So, $6 \times 7 = 42$

➤ Use facts with 5 and 2 to multiply by 7.
Break the 7 into a fact with 5 and a fact with 2.
To find 7×8:

Think: $5 \times 8 = 40$

$40 + 16 = 56$

$2 \times 8 = 16$

So, $7 \times 8 = 56$

➤ Use facts with 10
   to multiply by 9.
   To find 9 × 4, think:
   10 × 4 less 1 × 4
   10 × 4 = 40
   Subtract one group of 4.
   40 − 4 = 36
   9 × 4 = 36

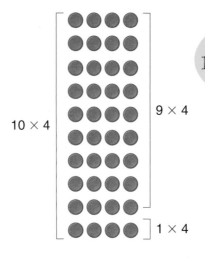

10 × 4   9 × 4

1 × 4

> When 9 is a factor,
> I think about multiplying
> by 10!

➤ Use a half then double,
   to multiply by an even factor.
   Here is another way to find 6 × 7:
   Choose the
   even factor, 6.
   Half of 6 is 3.
   Think of 3 × 7,
   then double.

> When I multiply by an
> even number, I think about
> how I can use doubles.

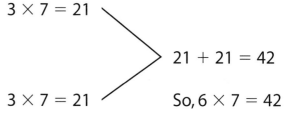

3 × 7 = 21

3 × 7 = 21

21 + 21 = 42

So, 6 × 7 = 42

Practice

1. Multiply. What strategies did you use?
   a) 3 × 6 **b)** 6 × 3 **c)** 7 × 5 **d)** 5 × 7
   e) 3 × 3 **f)** 6 × 6 **g)** 7 × 7 **h)** 7 × 6

2. Multiply. What strategies did you use?
   a) 9 × 2 **b)** 9 × 4 **c)** 9 × 8 **d)** 9 × 5
   e) 9 × 3 **f)** 9 × 6 **g)** 9 × 9 **h)** 9 × 7

3. Name two facts to help you find each product.
   a) 7 × 7 **b)** 3 × 8 **c)** 6 × 6 **d)** 8 × 6

4. Alexis is having a birthday party in 4 weeks.
   How many days does she have to wait?

5. You have 6 nickels and 4 dimes.
How many cents do you have?
Draw a picture.

6. How can you find the product of 9 × 7,
if you know the product of 9 × 5?
Show your work.

7. Rabia bought 4 bags of oranges.
Each bag had 6 oranges.
How many oranges did Rabia buy?

8. Write a story problem using the information given below.
Write an equation for the problem.
Solve the problem.
Include a multiplication fact as part of your answer.
   a) There are 4 wheels on a wagon. There are 8 wagons.
   b) There are 2 wheels on a bicycle. There are 9 bicycles.
   c) There are 3 wheels on a tricycle. There are 7 tricycles.

9. Write a story problem for this question: 6 × 7
Solve your problem. Show your work.

10. Without multiplying, how do you know that the product
of 6 × 5 is less than the product of 7 × 6?

11. Write a story problem for this equation: 7 × 8 = ☐
Solve the equation to solve the problem.

At Home

Reflect

You have learned different
ways to multiply.
Which way do you prefer?
Why?

Ask family members about their
strategies for remembering
multiplication facts.
How do their strategies compare
with yours?

Using Patterns in a Multiplication Chart

The **row** and **column** for the same factor have the same numbers.
The factors 6 and 7 are highlighted.
They show that:
$6 \times 7 = 42$ and $7 \times 6 = 42$

Column

x	1	2	3	4	5	6	7
1	1	2	3	4	5	6	7
2	2	4	6	8	10	12	14
3	3	6	9	12	15	18	21
4	4	8	12	16	20	24	28
5	5	10	15	20	25	30	35
6	6	12	18	24	30	36	42
7	7	14	21	28	35	42	49

← Row

Explore

Look at this multiplication chart.
How is it the same as the chart above? How is it different?
Use the chart to write ten multiplication facts.

x	0	1	2	3	4	5	6	7	8	9
0	0	0	0	0	0	0	0	0	0	0
1	0	1	2	3	4	5	6	7	8	9
2	0	2	4	6	8	10	12	14	16	18
3	0	3	6	9	12	15	18	21	24	27
4	0	4	8	12	16	20	24	28	32	36
5	0	5	10	15	20	25	30	35	40	45
6	0	6	12	18	24	30	36	42	48	54
7	0	7	14	21	28	35	42	49	56	63
8	0	8	16	24	32	40	48	56	64	72
9	0	9	18	27	36	45	54	63	72	81

Show *and* Share

Share your facts with another pair of students.
What patterns do you see in products with factors of 2?
With factors of 5?
What other patterns can you find?

Connect

You can use patterns to remember multiplication facts.

➤ In a multiplication chart, there are matching numbers on each side of the diagonal from 0 to 81.

You can use these numbers to help you remember multiplication facts.

x	0	1	2	3	4	5	6	7	8	9
0	0	0	0	0	0	0	0	0	0	0
1	0	1	2	3	4	5	6	7	8	9
2	0	2	4	6	8	10	12	14	16	18
3	0	3	6	9	12	15	18	21	24	27
4	0	4	8	12	16	20	24	28	32	36
5	0	5	10	15	20	25	30	35	40	45
6	0	6	12	18	24	30	36	42	48	54
7	0	7	14	21	28	35	42	49	56	63
8	0	8	16	24	32	40	48	56	64	72
9	0	9	18	27	36	45	54	63	72	81

Try to remember as many facts as you can.

If you know:	then you know:
$7 \times 6 = 42$	$6 \times 7 = 42$
$8 \times 5 = 40$	$5 \times 8 = 40$
$9 \times 4 = 36$	$4 \times 9 = 36$

➤ You can use patterns to remember multiplication facts with 9.

The number multiplied by 9 is always 1 more than the tens digit in the product; for example:

6 is 1 more than 5. ⟶

7 is 1 more than 6. ⟶

$1 \times 9 = 9$

$2 \times 9 = 18$

$3 \times 9 = 27$

$4 \times 9 = 36$ ⟶

$5 \times 9 = 45$ ⟶

$6 \times 9 = 54$

$7 \times 9 = 63$

$8 \times 9 = 72$

$9 \times 9 = 81$

The digits in the product always add to 9; for example:

⟶ $3 + 6 = 9$

⟶ $4 + 5 = 9$

1. What are the missing numbers?
   a) $7 \times 6 = \square \times 7$ b) $8 \times 3 = 3 \times \square$ c) $\square \times 9 = 9 \times 6$ d) $4 \times \square = 6 \times 4$
   Explain how you know.

2. Multiply. Which strategies did you use?
   a) 5×8 b) 3×9 c) 4×4 d) 6×7 e) 2×7
   f) 9×5 g) 8×8 h) 8×6 i) 1×7 j) 4×7

3. How can you use patterns to find each product?
   a) 7×9 b) 8×9 c) 9×9 d) 9×6 e) 5×9

4. If you know 7×8, what else do you know?

5. Show how you know each product.
   a) 8×5 b) 8×6 c) 6×9 d) 7×8 e) 6×3

6. How many days are there in 8 weeks? 9 weeks? How do you know?

7. Write a multiplication fact for each product.
   How many different facts can you find for each product?
   a) 12   b) 16   c) 18   d) 24   e) 36

8. Use a copy of the multiplication chart.
   Colour a design on the chart.
   Write a multiplication fact for each product you coloured.
   Exchange your facts with a classmate.
   Draw your classmate's design on another multiplication chart.

9. Yana walks his dog every day for 2 hours.
   How many hours does Yana walk his dog in 5 weeks?
   Show your work.

Reflect

A student cannot remember that $9 \times 7 = 63$.
What strategy might the student use to remember this fact?
Use words, pictures, or numbers to explain.

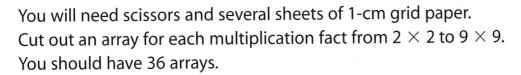

Array, Array!

You will need scissors and several sheets of 1-cm grid paper.
Cut out an array for each multiplication fact from 2 × 2 to 9 × 9.
You should have 36 arrays.

For each array:

- Write the products of factors on one side.

$$8 \times 2$$
$$2 \times 8$$

- Write the product on the other side.

16

Game 1 Matching Arrays

➤ Spread out the arrays.
18 cards should have grid side up.
18 cards should have grid side down.

➤ Take turns to choose an array and say what is on its other side.
If you are correct, you keep the array.
If you are incorrect, put the array back on the table.

➤ The winner is the student with more arrays at the end of the game.

Game 2 Who Has the Greater Product?

➤ Deal the cards, grid side up.

➤ Each student places one card on the table.
The student with the greater product takes both cards.

➤ Subtract the products.
The answer is the number of points the student gets.
Use a tally chart to keep score.

➤ The winner is the student with more points at the end of the game.

Strategies Toolkit

Explore

Mrs. Chan has triangular tables in her library.
She arranges the tables into one long row.
The tables fit together as shown.
One person can sit at each side of a table.
Mrs. Chan needs to seat 25 people.
How many tables does she need?

Show *and* Share

Describe the strategy you used to solve the problem.

Connect

Mr. Pasma has to seat 32 people at square tables.
He arranges the tables into one long row.
One person can sit at each side of a table.
How many tables does Mr. Pasma need?

Strategies

- **Make a table.**
- **Use a model.**
- **Draw a picture.**
- **Solve a simpler problem.**
- **Work backward.**
- **Guess and test.**
- **Make an organized list.**
- **Use a pattern.**

Understand

Plan

What do you know?
- The tables are square.
- There is a maximum of 4 seats at each table.
- The tables are arranged in a row.

Think of a strategy to help you solve the problem.
- You can **use a pattern**.
- Use orange Pattern Blocks to model the tables.
- List the numbers of tables and the numbers of seats.

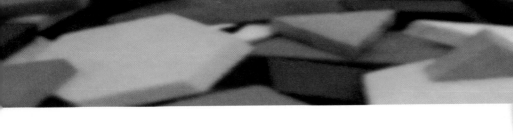
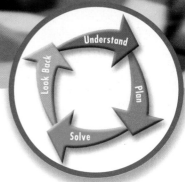

Record your list.

Number of Square Tables	Number of Seats
1	4
2	6

Look for patterns.
Continue the patterns to find
the number of tables needed to seat 32 people.

Check your work.
Does your answer make sense? Explain.

Practice

Choose one of the

Strategies

1. Suppose you have regular hexagonal tables.
   You want to seat 42 people.
   The tables will be joined in a row.
   How many tables do you need?

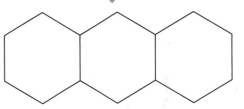

2. Pool decks come in many shapes and sizes.
   Use grid paper to model this pattern for a deck.
   How many blue tiles are there in the figure that has 20 red tiles?

   Figure 1   Figure 2   Figure 3

Reflect

How can you use a pattern to solve a problem?
Use words and numbers to explain.

Using Arrays to Divide

You used multiplication facts to describe an array.

3 rows of 4 counters
$3 \times 4 = 12$

Explain how the counters are divided into rows.
Write the division fact.

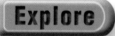 **Explore**

You will need counters.
There are 24 children in the parade.
They have to line up in equal rows.
How many children could be in each row?
How many different ways can you find?
Use counters to model each way.
Record each model.

Show *and* Share

Share your answers with another pair of students.
How many different ways can the children line up in equal rows?
Write a division fact for each way.

There are 12 drum dancers.
They will dance onto the stage in equal rows.
How many dancers could be in each row?

We can use arrays to show the possible ways.

2 rows of 6
12 ÷ 2 = 6

4 rows of 3
12 ÷ 4 = 3

6 rows of 2
12 ÷ 6 = 2

3 rows of 4
12 ÷ 3 = 4

1 row of 12
12 ÷ 1 = 12

12 rows of 1
12 ÷ 12 = 1

Practice

Use counters when they help.

1. Write a division fact for each array.

a)

b)

c)

2. a) Draw an array to show 5 ÷ 1.
  b) Draw an array to show 5 ÷ 5.

3. Divide.
  a) 6 ÷ 1   **b)** 8 ÷ 1   **c)** 7 ÷ 1   **d)** 1 ÷ 1
  What patterns do you see?

4. Copy and complete each division equation.
  a) 30 ÷ 6 = □   **b)** 7 ÷ □ = 7   **c)** □ ÷ 7 = 4   **d)** 36 ÷ 6 = □

5. The choir sings on stage.
  a) There are 35 chairs in 5 equal rows.
     How many chairs are there in each row?
  b) Suppose there are 35 chairs in 7 equal rows.
     How many chairs are there in each row?
  Draw an array to show each answer.

6. There are 12 drummers and 15 horn players.
  a) Can they form equal rows of 2?
     How do you know?
  b) Can they form equal rows of 3?
     How do you know?
  c) What other equal rows can they form?
  Show your work.

7. Explain why the answer is always 1 when you divide a number by itself.

8. Write a story problem that you can solve by drawing an array to divide.
  Solve your problem.
  Show your work.

Reflect

How can you use an array to divide?
Use words, numbers, or pictures to explain.

Relating Multiplication and Division

Explore

Fifty-six students want to play basketball.
There are 7 players on a school team.
How many teams can be made?
Solve this problem.
Use any materials you need.
Record your work.

Show and Share

Share your answer with another pair of students.
What strategy did you use to solve the problem?

Connect

Thirty students want to play volleyball.
There are 6 players on a team.
How many teams can there be?

To find how many teams, divide: $30 \div 6$

➤ Make an array of 30 counters with 6 counters in each row.
There are 5 rows.

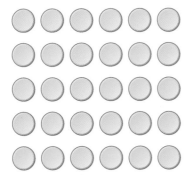

So, $30 \div 6 = 5$
There can be 5 teams.

➤ To find 30 ÷ 6:

To divide, you can think about multiplication.

Think: 6 times which number is 30?

You know 6 × 5 = 30.
So, 30 ÷ 6 = 5

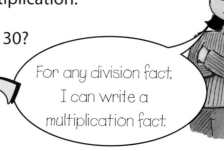

For any division fact, I can write a multiplication fact.

Practice

Use counters when they help.

1. Write a multiplication fact and a division fact for each array.

a)

b)

c)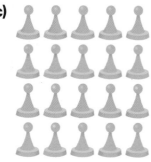

d)

2. a) Draw an array to show 7 ÷ 7.
   b) Draw an array to show 7 ÷ 1.

3. Use multiplication facts to help you divide.

a) 6 × 4 = 24   b) 5 × 7 = 35   c) 4 × 2 = 8   d) 9 × 6 = 54
   24 ÷ 4 = ☐   35 ÷ 7 = ☐   8 ÷ 2 = ☐   54 ÷ 6 = ☐

4. Divide. Draw a picture to show your work.
   a) 3 ÷ 1   **b)** 14 ÷ 2   **c)** 20 ÷ 4   **d)** 42 ÷ 7
   e) 25 ÷ 5   **f)** 24 ÷ 6   **g)** 21 ÷ 3   **h)** 6 ÷ 6

5. Danielle works in a hardware store.
   She makes packages of washers.
   Danielle puts 6 washers in each package.
   She has 36 washers.
   How many packages can Danielle make?

6. Use a multiplication fact to divide.
   a) 15 ÷ 3   **b)** 30 ÷ 5   **c)** 49 ÷ 7   **d)** 12 ÷ 2
   e) 28 ÷ 4   **f)** 42 ÷ 6   **g)** 6 ÷ 1   **h)** 16 ÷ 4

7. Write a story problem you can solve using division.
   Trade problems with a classmate.
   Solve your classmate's problem.

8. A class made equal teams of 5 for basketball,
   and equal teams of 6 for volleyball.
   Each student in the class was on a team.
   How many students might be in the class?

9. Joe has 35 cubes.
   He shares the cubes equally among 7 students.
   Each student needs 6 cubes.
   Does Joe have enough cubes? Explain.
   Show your work.

Reflect

How can you use what you have learned
about multiplication to divide?
Use words, numbers, or pictures to explain.

9

Dividing by Numbers from 1 to 9

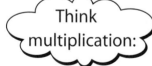

Use the multiplication chart.

➤ Write the multiplication facts that have 8 as a factor.
Use these facts to write all the division facts where you divide by 8.
Draw arrays to show some of these facts.

➤ Repeat the activity for multiplication facts that have 9 as a factor.

x	1	2	3	4	5	6	7	8	9
1	1	2	3	4	5	6	7	8	9
2	2	4	6	8	10	12	14	16	18
3	3	6	9	12	15	18	21	24	27
4	4	8	12	16	20	24	28	32	36
5	5	10	15	20	25	30	35	40	45
6	6	12	18	24	30	36	42	48	54
7	7	14	21	28	35	42	49	56	63
8	8	16	24	32	40	48	56	64	72
9	9	18	27	36	45	54	63	72	81

Show *and* Share

Share your facts and arrays with another pair of students.
How do you know if you found all the facts?

Connect

➤ To find 72 ÷ 9:

Think multiplication: $9 \times \square = 72$

You know: $9 \times 8 = 72$
So, $72 \div 9 = 8$
Also, $72 \div 8 = 9$

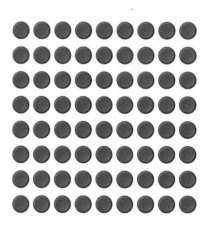

➤ To find 64 ÷ 8:

Think multiplication: $8 \times \square = 64$

You know: $8 \times 8 = 64$
So, $64 \div 8 = 8$

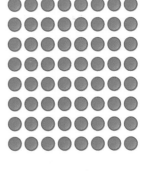

For most multiplication facts, you know two division facts.

For some multiplication facts, you know one division fact.

$7 \times 8 = 56$
$8 \times 7 = 56$ These are **related facts**.
$56 \div 8 = 7$
$56 \div 7 = 8$

$7 \times 7 = 49$
$49 \div 7 = 7$

Use counters or grid paper when they help.

Practice

1. Write two multiplication facts and two division facts for each array.

   a)

   b)

2. Find each product.
   Then write a related multiplication fact and two division facts.
   a) $7 \times 3 = \square$ **b)** $8 \times 6 = \square$ **c)** $5 \times 9 = \square$ **d)** $9 \times 7 = \square$

3. Write four related facts for each set of numbers.
   a) 9, 4, 36   **b)** 5, 8, 40   **c)** 4, 7, 28   **d)** 1, 7, 7

4. **a)** One number in a set of related facts is 63. What could the facts be?
   b) One number in a set of related facts is 8. What could the facts be?

5. Divide.
   a) 24 ÷ 8   **b)** 36 ÷ 9   **c)** 56 ÷ 8   **d)** 9 ÷ 1   **e)** 16 ÷ 8
   f) 72 ÷ 8   **g)** 63 ÷ 9   **h)** 27 ÷ 3   **i)** 8 ÷ 8   **j)** 64 ÷ 8

6. Write all the multiplication facts for which there is
   only one division fact.
   Where are these products on the multiplication chart?
   Write each related division fact.

7. Divide.
   a) 35 ÷ 5   **b)** 45 ÷ 9   **c)** 81 ÷ 9   **d)** 18 ÷ 6
   e) 20 ÷ 4   **f)** 36 ÷ 9   **g)** 54 ÷ 9   **h)** 63 ÷ 9

8. **a)** If you know that 63 ÷ 9 = 7, what else do you know?
   b) Write a story problem that could be represented by the equation in part a.

9. Grade 4 students are going on an activities day.
   There are 32 students in the class.
   Eight students can go in each canoe.
   How many canoes will be needed?
   Write an equation for this problem.
   Solve the equation.

10. There are 9 marbles in each bag.
    Heidi wants to buy 54 marbles.
    How many bags does Heidi need to buy?

11. Write a story problem that you can solve by dividing.
    Trade problems with a classmate. Solve your classmate's problem.

12. **a)** Is 48 ÷ 8 more or less than 40 ÷ 8? How do you know?
    b) Is 72 ÷ 9 more or less than 72 ÷ 8? How do you know?
    Show your work.

Reflect

How can you use an array to show how multiplication
and division are related?

Pose and Solve Problems

Explore ....

Thirty children signed up for sports.
This table shows the sport and the number of players per team.

Sport	Number of Players
baseball	9
basketball	5
soccer	6
frisbee	8

The coaches want every child to be on a team.
The children play one sport at a time.
Which sports can the coaches choose?

Solve this problem.
Use any materials you need.
Record your work.

Show *and* Share

Share your work with another pair of students.
Did you multiply or divide to solve the problem?
What strategy did you use to solve the problem?

In a tennis match, players play
doubles games or singles games.
4 players are needed for a doubles game.
2 players are needed for a singles game.
There are 22 players and 7 games.
How many doubles games
and singles games are there?

Here are two ways to find out.

➤ Use a model.
Use 22 counters to show the number of players.
Put the counters in groups of 2 to show singles games.
Put the counters in groups of 4 to show doubles games.
Make sure there are 7 groups.

$$4 \times 4 = 16 \qquad 3 \times 2 = 6$$

$$16 + 6 = 22$$

➤ Guess, then test.
Suppose you guess 2 doubles games and 5 singles games.
$2 \times 4 = 8$; that is 8 players playing doubles.
$5 \times 2 = 10$; that is 10 players playing singles.
Test: $8 + 10 = 18$; that is too few players.

Try another guess.
Guess 4 doubles games and 3 singles games.
$4 \times 4 = 16$; that is 16 players playing doubles.
$3 \times 2 = 6$; that is 6 players playing singles.
Test: $16 + 6 = 22$; yes, 22 is the correct number of players.

There are 4 doubles games and 3 singles games.

Use counters if they help. Show your work.

1. Forty-two students want to play volleyball.
   There are 6 players on a team.
   How many teams can there be?

2. There should be a water bottle for each player in the tournament.
   There are 8 teams. Each team has 6 players.
   How many water bottles are needed?

3. Write a story problem for each situation. Then solve the problem.
   a) There are 24 water bottles for 4 teams.
   b) There are 8 teams with 9 players on each team.

4. The coach has between 40 and 50 ribbons for the track meet.
   She has an equal number of ribbons for each of the 6 events.
   There are 5 ribbons left over.
   How many ribbons might there be for each event?

5. Thirty-five students signed up to play hockey.
   There can be 7 teams in the tournament.
   Each team should have 6 players.
   Are there enough players to make 7 teams? How do you know?

6. Use the data in the table.
   Write 3 story problems you can solve
   using multiplication or division.
   Solve each problem.

Sport	Players on a Team
Baseball	9
Basketball	5
Ice hockey	6

7. Write a story problem you could solve
   by finding 9×5.

Reflect

How do you choose a strategy to solve a story problem?
Which strategy do you use often? Explain.

LESSON

Use any materials when they help.

1   **1.** Multiply. What strategies did you use?

a) 3×2 b) 3×4 c) 3×8

d) 2×6 e) 4×6 f) 8×6

2 **4**   **2.** Multiply. What strategies did you use?

a) 5×8 b) 1×8 c) 9×0 d) 7×6

e) 9×9 f) 8×10 g) 8×9 h) 4×8

3. There are 8 cereal bars in a box.

a) How many bars are in 2 boxes?

b) How many bars are in 4 boxes?

c) How many bars are in 8 boxes?

How could you use one answer to get the next?

3   **4.** Copy this chart. Fill in the missing numbers.
Describe each skip counting pattern.

×	4	5	6	7	8	9
4	16	20				
5		25		35		
6			36			54

5. Ali knows that $7 \times 6 = 42$.
How can he use this fact to find 7×7 and 7×8?

5   **6.** The answer to a multiplication question is 24.
What might the question be?
Write as many multiplication facts as you can.

7. Use words, numbers, or pictures to explain your thinking.

a) How might you use 4×7 to find 8×7?

b) How might you use 5×5 to find 5×6?

c) How might you use 3×10 to find 3×9?

8. Write a multiplication fact and a division fact for each array.

a)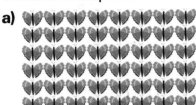

b)

9. What might each missing number be? How many answers can you find?

a) $\square \times \bigcirc = 35$ b) $\triangle \div \square = 1$ c) $\square \times \bigcirc = 63$ d) $\square \div \triangle = 5$

10. Divide.

a) $45 \div 9$ b) $32 \div 8$ c) $56 \div 7$ d) $27 \div 3$ e) $9 \div 9$

11. Find each product.
Then write a related multiplication fact and two division facts.

a) $6 \times 7 = \square$ b) $6 \times 9 = \square$ c) $8 \times 3 = \square$ d) $5 \times 7 = \square$

12. Write four related facts for each set
of numbers.

a) 9, 7, 63 b) 4, 5, 20 c) 6, 8, 48

13. Which multiplication facts will help you
find the answers?

a) $30 \div 6$ b) $64 \div 8$ c) $42 \div 7$

14. There are 8 flowers in a bunch.
Suppose you want to buy 40 flowers.
How many bunches would you need to buy?
Write an equation for the problem.
Solve the equation.

15. Use the data below.
Write a problem that can be solved by
multiplying or dividing.
Solve your problem.

There are 48 apples.
There are 5 boys and 3 girls.

UNIT

3 Learning Goals

☑ use different mental math strategies to multiply and divide

☑ multiply by 0, 1, and 10

☑ divide by 1

☑ recall multiplication and division facts

☑ identify and describe patterns in a multiplication chart

☑ relate multiplication and division

☑ pose and solve story problems using multiplication and division

☑ write and solve equations

Here Comes the Band!

Part 1

A marching band has 48 people.
They march in equal rows.
How many different ways can the band be arranged?
Write a multiplication fact for each way.
Show your work.

Part 2

➤ Thirty band members perform onstage.
You will set up chairs for them.
- There must be equal rows of chairs.
- There must be at least 2 chairs
  in each row.

How many ways can you set up the chairs?
Write an equation you can solve to find
each way.

➤ What problems might you have
if the band has 31 members? Explain.

Part 3

Suppose you are the bandleader for a day.
- You choose how many band members
  will play that day.
- Choose a number.
- Show all the different ways
  you could arrange your band members
  into equal rows.
  Write a multiplication fact and
  a division fact for each way.

Check List

Your work should show
- ☑ all the ways to arrange
  the band members in
  equal rows
- ☑ correct multiplication and
  division facts
- ☑ a clear explanation of why
  31 band members might
  be a problem
- ☑ how you used
  mathematical language
  and symbols correctly

Reflect on Your Learning

How are multiplication and division related?
Use words, pictures, or numbers to explain.

UNIT
1

1. What patterns do you see in each set of numbers?
   a) the blue numbers
   b) the red numbers
   c) the green numbers

×	0	1	2	3	4	5	6	7	8	9
0	0	0	0	0	0	0	0	0	0	0
1	0	1	2	3	4	5	6	7	8	9
2	0	2	4	6	8	10	12	14	16	18
3	0	3	6	9	12	15	18	21	24	27
4	0	4	8	12	16	20	24	28	32	36
5	0	5	10	15	20	25	30	35	40	45
6	0	6	12	18	24	30	36	42	48	54
7	0	7	14	21	28	35	42	49	56	63
8	0	8	16	24	32	40	48	56	64	72
9	0	9	18	27	36	45	54	63	72	81

2. Use the multiplication chart in question 1.
   Find 3 patterns different from those in question 1.
   Write a rule for each pattern.

3. Find the missing numbers in this addition chart.
   Explain how you could use patterns to find these numbers.

+	4	8	12	16	20
2	6		14	18	
4	8	12		20	24
6	10		18		26
8	12	16		24	
10		18		26	30

4. Find the errors in this multiplication chart.
   How did you find each error?

×	10	8	6	4	2
2	20	16	12	8	6
4	40	30	24	16	8
6	60	48	36	26	12
8	80	64	50	32	16
10	100	80	60	40	20

5. Choose a 4-digit number. Write it in:
  a) standard form   **b)** expanded form
  c) words   **d)** a place-value chart

6. Use the digits 3, 4, 5, 6.
  a) Write all the 4-digit numbers greater
     than 4000 and less than 5000.
  b) Order the numbers from greatest to least.
     Use place value to explain how you did this.

7. Estimate first. Then add or subtract the numbers
   for which the answer is greater than 500.
   a)   219   **b)**   627   **c)**   87   **d)**   786
       − 195   + 186   + 256   − 195

8. Multiply. Use mental math.
   Try to use a different strategy each time.
   a) 6 × 3   **b)** 6 × 6   **c)** 6 × 0   **d)** 2 × 7   **e)** 4 × 7
   f) 5 × 7   **g)** 7 × 7   **h)** 8 × 9   **i)** 4 × 1   **j)** 5 × 9

9. Megan has 8 boxes of books.
   Each box contains 8 books.
   How many books does Megan have altogether?

10. Divide. How can you use multiplication to help you?
   a) 20 ÷ 4   **b)** 42 ÷ 7   **c)** 5 ÷ 5   **d)** 27 ÷ 3   **e)** 16 ÷ 2
   f) 48 ÷ 8   **g)** 30 ÷ 6   **h)** 81 ÷ 9   **i)** 21 ÷ 7   **j)** 32 ÷ 4

11. Choose one part of question 10.
   Draw an array to show how to divide.

12. Ali has 56 apples to share equally among 7 people.
   How many apples does each person get?
   How do you know your answer is correct?

UNIT 4

Measurement

Design a Playground

Key Words

analog clock

digital clock

elapsed time

24-hour clock

units

area

square units

square centimetre

square metre

- What does the sign on the gate mean?
- Suppose you went to the playground at 4 o'clock and stayed for 1 hour. What time would you leave?
- Which section of the playground covers the most space?
- Which sections cover about the same amount of space?

123

Calendar Time

What do you know from looking at these calendar pages?

FEBRUARY 2004

S	M	T	W	T	F	S
1	2 Groundhog Day	3	4	5	6	7
8	9	10	11	12 Math Test	13	14 ♥
15	16	17 Dentist 10:00	18	19	20	21
22	23	24	25 Ben's Birthday	26	27	28
29						

FEBRUARY 2005

S	M	T	W	T	F	S
		1	2 Groundhog Day	3	4	5
6	7	8	9 Library	10	11	12
13	14 ♥	15	16	17	18	19
20	21	22	23	24	25 Ben's Birthday	26
27	28					

Explore

➤ Talk to 5 classmates.
  Find out the month, day of the month,
  and year each classmate was born.
  Record your results in a table.

➤ Write your own date of birth in the table.
  Write the dates of birth of some of your
  family members.

➤ Order your list of people from oldest to youngest.

Birthdays

Name	Month	Day	Year
Sindar	Nov.	19th	
Jacob			
Dallas			
Moira			
Saffa			

Show *and* Share

Share your list of birth dates with another pair of classmates.
Talk about how you recorded the dates.
Explain the strategy you used to order your list of people from oldest to youngest.

➤ The circled date is October 24th, 2006.

We can write this date in metric notation:

We start with the greatest unit of time – the year, and go to the least unit of time – the day.

2006	10	24
↑	↑	↑
Year	Month	Day

October is the 10th month.

➤ When we write the date in metric notation, we use 2 digits for the month and 2 digits for the day.

The circled date is April 8th, 1999.
In metric notation, we write:

We put a zero in the tens place for numbers from 1 to 9.

1999	04	08
↑	↑	↑
Year	4th month (April)	8th day of the month

➤ Some people also use 2 digits for the year.
Then, April 8th, 1999 could be written: 99 04 08

➤ Other ways to write dates are month, day, year and day, month, year.
This means that a date written as 05 04 01 could be interpreted two ways:

05	04	01	or	05	04	01
↑	↑	↑		↑	↑	↑
5th month	4th day	Year		5th day	4th month	Year
May 4th, 2001 or May 4, 2001				5 April, 2001		

1. Write each date in metric notation.
   a) December 14th, 2002   b) April 25, 2110
   c) May 5, 1941   d) October 1st, 2000
   e) September 8, 2005   f) June 25, 1927

2. Write each date using words and numbers.
   a) 2008 11 04   b) 2025 03 09
   c) 1846 08 13   d) 2007 10 19
   e) 2006 05 28   f) 2000 01 01

3. Use metric notation to record each circled date on the calendar at the right.

4. Write each date in metric notation.
   a) the first day of next month
   b) the last day of this month
   c) the date of your 15th birthday
   d) the last day of this year
   e) 5 months before October 31st of this year
   f) 2 months after today

5. Joe wrote his birth date as 02 08 94.
   Jane was confused.
   Did Joe mean February 8, 1994 or August 2, 1994?
   Joe said, "I was born in the winter."
   Write Joe's birth date in metric notation.

6. Read each statement.
   Re-write the date in metric notation.
   a) Nunavut entered confederation on April 1st, 1999.

b) The American spacecraft *Apollo 11* landed on the moon on July 20th, 1969.

c) Sir John A. MacDonald was the first Prime Minister of Canada.
He was born on January 11th, 1815.

d) The first Oreo cookie was sold on March 6th, 1912.

e) The rings of Uranus were discovered on March 10th, 1977.

f) The first toy ever advertised on TV was Mr. Potato Head, on April 30th, 1952.

g) Chewing gum was patented on December 28th, 1869.

7. Match each date to the correct calendar page below.

a) 2007 07 08   **b)** 2007 06 17   **c)** 2007 08 07

JUNE 2007						
S	M	T	W	T	F	S
					1	2
3	4	5	6	7	8	9
10	11	12	13	14	15	16
(17)	18	19	20	21	22	23
24	25	26	27	28	29	30

JULY 2007						
S	M	T	W	T	F	S
1	2	3	4	5	6	7
(8)	9	10	11	12	13	14
15	16	17	18	19	20	21
22	23	24	25	26	27	28
29	30	31				

AUGUST 2007						
S	M	T	W	T	F	S
			1	2	3	4
5	6	(7)	8	9	10	11
12	13	14	15	16	17	18
19	20	21	22	23	24	25
26	27	28	29	30	31	

8. Use the calendar pages above.
Write each date using words and numbers.

a) 19 08 2007   **b)** 27 06 2007   **c)** 11 07 2007

9. Dennis left for Yellowknife on 2006 07 24 and returned on 2006 08 05.
How long was Dennis away?
Show your work.

Reflect

Use metric notation to write a date that is important to you.
Tell why it is important.

Exploring Time

It's 8 o'clock.
Tyrell gets up.

It's quarter after 8.
Tyrell eats breakfast.

It's half past 8.
Tyrell leaves for school.

It's quarter to 9.
Tyrell arrives at school.

Explore

➤ Use the clock cards your teacher gives you.
Place the cards in a pile face down.
Take turns to turn over the top card.
Say the time the clock shows.
Continue until all the cards have been used.

➤ Choose a card from the pile.
Draw a picture to show what you might do at that time.
Repeat with two more cards.

Show *and* Share

Share your pictures with your partner.
Tell whether each picture happens in the morning, the afternoon, or at night.

Connect

➤ A clock with numbers and hands is an **analog clock**.
A clock face shows the numbers from 1 to 12.

➤ There are 24 hours in 1 day. Each day,
the hour hand moves twice around the clock.
It takes 1 hour for the hour hand to
move from one number to the next.

➤ There are 60 minutes in 1 hour.
Each hour, the minute hand moves once around the clock.

➤ It takes 15 minutes for the minute hand to move
$\frac{1}{4}$ of the way around the clock.
It is 15 minutes after 10 o'clock.
We say: "It is *quarter after* 10" or "ten fifteen."

➤ It takes 30 minutes for the
minute hand to move half
way around the clock.
It is 30 minutes
after 10 o'clock.
We say: "It is *half past* 10"
or "ten thirty."

The hour hand is
half way between the
10 and the 11.

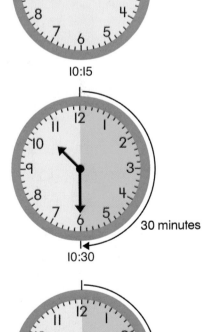

➤ It takes 45 minutes for
the minute hand to
move $\frac{3}{4}$ of the way
around the clock.
It is 45 minutes
after 10 o'clock.
We say: "It is ten forty-five."

When the minute
hand has moved
through another $\frac{1}{4}$ turn, it
will be 11 o'clock. So,
we can also say: "It is
quarter to 11."

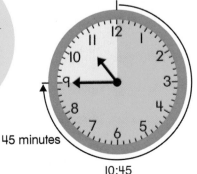

➤ A clock with numbers and no hands is a **digital clock**.
It shows the time using numbers and a colon.

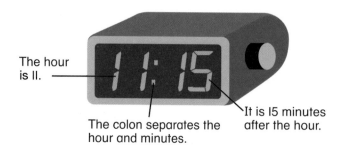

The hour is II.

The colon separates the hour and minutes.

It is 15 minutes after the hour.

The clock shows 15 minutes after 11 o'clock. We say: "Eleven fifteen"

Practice

1. Match each analog clock with the digital clock that shows the same time.

a)
b)
c)
d)

A
B
C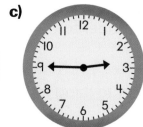
D

2. Write each time in two ways. The first one is done for you.

a)
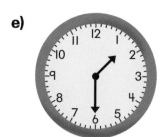
12:30
half past 12

b)

c)

d)

e)

f)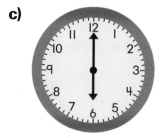

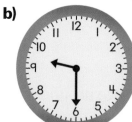
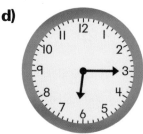
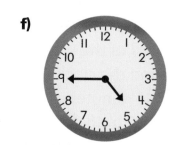

130

3. Write each time in a different way.

    a) 6:00   **b)** quarter to 1   **c)** eleven forty-five   **d)** 4:15

4. Draw a digital clock to show each time.

    a) seven thirty   **b)** one fifteen   **c)** four forty-five

    d) three o'clock   **e)** quarter to eight **f)** half past two

5. Draw and label a picture to show what you might be doing at each time.

    a) 5:00 in the morning   **b)** 2:45 in the afternoon

6. Jodie started her homework at 5:00. She worked for $\frac{1}{2}$ an hour.
At what time did Jodie finish?

7. Ho went fishing for 45 minutes. He started at quarter to 7.
At what time did Ho finish?

8. Stefan looked at this clock and said, "It is quarter to six."
Petra looked at the clock and said, "It is five forty-five."
Who is correct? Explain.

9. Jessie practised archery for half an hour. She started at 3:30.
At what time did she finish?

10. Julia's school has two 15-minute recesses each day –
one in the morning and one in the afternoon.

    a) What time might each recess start and end?

    b) Draw an analog clock to show each start time.

    c) Draw a digital clock to show each end time.

Reflect

Which clock do you prefer to use – analog or digital?
Explain your choice.

Telling Time

It takes 5 minutes for the minute hand to move from one number to the next number.

5 minutes after 3 o'clock

10 minutes after 3 o'clock

Explore

➤ Use the 9 clock cards your teacher gives you.
Make a set of 9 time cards to match the clock cards.

7:25

➤ Mix the time cards and the clock cards.
Trade with your partner.
Sort your partner's cards into 9 matching sets.

6:50

Show and Share

Check each other's work.
How did you match the cards?

Math Link

Number Sense

You can multiply by 5 to find the time in minutes.

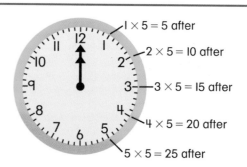

$1 \times 5 = 5$ after
$2 \times 5 = 10$ after
$3 \times 5 = 15$ after
$4 \times 5 = 20$ after
$5 \times 5 = 25$ after

➤ This analog clock shows
20 minutes after 9 o'clock.

We write: 9:20
We say: "Twenty after nine"
or "Twenty past nine"
or "Nine twenty"

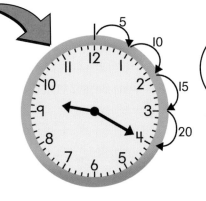

Skip count
the minutes:
5, 10, 15, 20

➤ This analog clock shows
55 minutes after 11 o'clock or
5 minutes before 12 o'clock.

We write: 11:55
We say: "Five before 12"
or "Five to twelve"
or "Eleven fifty-five"

Skip count the minutes:
5, 10, 15, 20, 25, 30, 35,
40, 45, 50, 55

➤ This digital clock shows
20 minutes after 4 o'clock.
We write: 4:20
We say: "Four twenty"

➤ This digital clock shows
5 minutes after 6 o'clock.
We write: 6:05
We say: "Six O five"

Practice

· Unit 4 Lesson 3

Use a play clock when it helps.

1. Write the time shown on each analog clock.
Skip count the minutes if you need to.

a)    **b)**    **c)**

d)    **e)**    **f)**

2. Write each time two ways. The first one is done for you.

a)  10 after 10
10:10

b)    **c)**

d)    **e)**    **f)**

3. Match each analog clock with the digital clock
that shows the same time.

a)    **b)**    **c)**    **d)**

A    **B**    **C**    **D**

4. a) Suppose it is 6:20.
What time will it be in 5 minutes?

b) Suppose it is 9:00.
What time will it be in 10 minutes?

c) Suppose it is 4:55.
What time will it be in 15 minutes?

5. School starts at 9:00.
   a) Corrina was 5 minutes late.
   Draw a digital clock to show what
   time she arrived.
   b) Sammy was 10 minutes early.
   Draw an analog clock to show what time he arrived.
   Write the time two ways.
   Show your work.

6. On an analog clock, show 10 minutes to 11.
Then write 10 minutes to 11 another way.

7. Pilan began to read at 2 o'clock.
He read for 60 minutes.
At what time did Pilan finish reading?

8. Draw a digital clock to show each time.
   a) eleven thirty-five
   b) half past two
   c) two twenty
   d) six fifty
   e) quarter to eleven
   f) one fifty-five

At Home

Reflect

When you tell time from an analog
clock, how do you know what the
hour is? Use words and pictures
to explain.

Find all the clocks in
your home.
Draw a picture of each
clock at a different time
during the day. Write each
time in words.

Elapsed Time

Explore

➤ Saba sent this invitation to 5 friends.
Help Saba plan the afternoon.
Choose 4 games to play.

➤ Make a chart to show each game
and its start and end times.
Remember to include a time for eating.

➤ Trade charts with another pair
of classmates.
Find how many minutes will be
spent playing each game.

Show *and* Share

Check each other's work.
Talk about the strategies you used
to figure out the time in minutes.

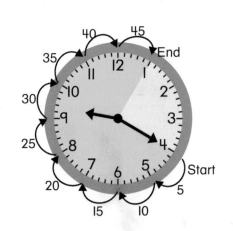

You are invited
to an afternoon
of fun & games

Place: 235 Hickory Hollow Drive
Date: Saturday, February 6
Time: 1:15 pm - 4:00 pm.

Food will be served

A.M.—times from midnight to noon
P.M.—times from noon to midnight

Connect

The amount of time from the start to the
end of an activity is the **elapsed time**.

Fatima and Clara played checkers from
9:20 A.M. to 10:05 A.M.

To find the elapsed time in minutes,
count on by 5s.

Fatima and Clara played checkers for 45 minutes.

Use a play clock when it helps.

1. Recess starts at 10:10 A.M. and ends at 10:25 A.M.
How long is recess?

2. Find each elapsed time.
   a) 9:15 A.M. to 10:10 A.M.
   b) 3:30 P.M. to 4:15 P.M.
   c) 11:50 A.M. to 12:10 P.M.

3. Aaron listened to music for 35 minutes.
He started at 6:25 P.M.
At what time did Aaron stop?

4. You have 20 minutes to clean your room.
You start at 3:45 P.M.
You finish at 4:10 P.M.
Did you clean your room in time? Explain.

5. Suppose it is now 7:50 P.M.
How many minutes will it be until 8:15 P.M.?
How do you know?

6. A spider took 30 minutes to spin its web.
The spider finished spinning at 11:40 A.M.
At what time did it start?

7. This chart shows Emma's Saturday activities.
Copy and complete the chart.

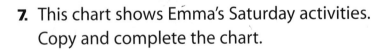

	Activity	Start Time	End Time	Elapsed Time
a)	Library visit	9:15 A.M.	9:55 A.M.	
b)	Hockey practice	4:25 P.M.		40 minutes
c)	Help with supper		7:10 P.M.	35 minutes

8. Arlo and Wilfred met at the lake at 2:30 P.M.
Arlo had taken 20 minutes to get to the lake.
Wilfred had taken 25 minutes.
At what time did each boy leave home? How do you know?

9. Look at each analog clock. What time will it be 45 minutes later?

a)  **b)** 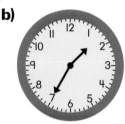 **c)** 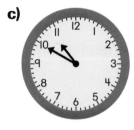 **d)**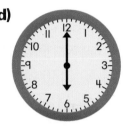

10. Here is a schedule of activities for the trip to the Outdoor Nature Centre.

Activity	Time Period
Snowshoeing	1:00 – 1:50
Bird-watching	2:00 – 2:30
Snack time	2:40 – 2:55
Snow sculpture	3:00 – 3:55

a) Why does the chart not show A.M. or P.M.?

b) Which activity takes the most time?
How much time does it take?

c) How much time do bird-watching and snow sculpture take together?

d) It takes 40 minutes to ride from school to the nature centre.
At what time should the class leave school?

e) About what time will the class arrive back at school after the trip? How do you know?

f) Make up your own question about the trip.
Answer your question.
Show your work.

Reflect

Name an activity you usually do before 9:00 A.M.
Name an activity you might do between 1:00 P.M. and 7:00 P.M.

Telling Time to the Minute

Maria can do 25 push-ups in 1 minute.
What can you do in 1 minute?

Explore

 Game

Make a set of 6 clock cards.
The minute hand on each clock should be between the 5-minute marks.

Now make a set of 6 time cards to match the times on the clock cards.

Play the Matching Time Game.

➤ Mix up the 12 cards.
Place the cards face down
in 3 rows of 4.

➤ Take turns turning over 2 cards.

➤ If the cards match, keep them.

➤ If the cards do not match,
put them back, face down.

➤ Keep playing until all the cards
have gone.

Show and Share

Talk about the strategies you used
to find matching pairs of cards.

It takes 1 minute for the minute hand to move from one mark
on the clock to the next mark.

20 minutes
after
7 o'clock

7:20

21 minutes
after
7 o'clock

7:21

You can read times after the half-hour in two ways.

We say:
54 minutes
after
11 o'clock
or 11:54

We say:
6 minutes before
12 o'clock
or 6 minutes to 12

Practice

1. Write the time shown on each clock.

a)

b)

c)

d)

e)

f)

2. Write each time two ways. Write A.M. or P.M.

a) school ends

b) dinner time

c) wake up

d) bed time

e) play baseball

f) lunch time

3. Draw an analog clock to show each time.

a) 9 minutes after 3   **b)** 14 minutes to 2   **c)** 27 minutes after 7

4. Draw a digital clock to show each time.

a) 14 minutes after 3   **b)** 8 minutes to 7

c) 36 minutes after 5   **d)** 7 minutes to 1

e) 25 minutes to 8   **f)** 10 minutes to 9

5. Bernard and Stephan agreed to meet at the music store.
Bernard arrived at 4:47 P.M.
Stephan arrived at quarter to 5.
Who arrived first?
How many minutes earlier was he than the other boy?
Show your work.

At Home

Reflect

Are 3:48 and 12 minutes to 4
the same time?
Use words and pictures to explain.

How is your sense of time?
Guess what time you think it
is right now.
Go and check a clock.
How close was your guess?

The 24-Hour Clock

What time is shown on the clock?
Can you tell from the clock if it is morning or afternoon?
Suppose a clock face had a different number for each hour of the day. What might the clock look like?

Explore

Planes leave Vancouver airport at the times shown.
They fly to the cities shown.

01:15 Hong Kong	15:15 Hong Kong
06:00 Toronto	20:50 London
11:10 Toronto	23:45 Toronto
14:55 London	

➤ Which flights leave very early in the morning? How do you know?

➤ Which flights leave before noon?

➤ Which flights leave after noon? How do you know?

➤ Which flights leave late at night? How do you know?

Show *and* Share

Share your results with another pair of students.
If the answers do not agree, try to find out who is correct.
Why do you think the times are written in this way
instead of using A.M. and P.M.?

Your friend says she will be at your house at 8 o'clock.
You need to know if she means 8 A.M. or 8 P.M.
There is another way to write the time where
we do not use A.M. or P.M. We use a **24-hour clock**.

➤ There are 24 hours in one day.
From midnight to noon, the hours are from 0 to 12.
From 1 o'clock to midnight, the hours are from 13 to 24.
When we use the 24-hour clock, we use 4 digits
to write the time.

9:45 A.M. is written 09:45.

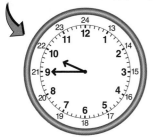

6:15 P.M. is written 18:15.

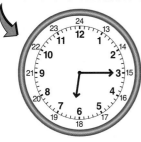

Before 10:00 A.M.,
I write 0 in front of
the hour. After noon, I add 12
to the hour.
So, 2 o'clock in the afternoon
is 12:00 + 2:00 = 14:00

➤ Kathy arrived at the library at 11:45 and left at 14:20.
How long did she spend in the library?
Count on to find the time.

We write 15 minutes as 15 min. We write 2 hours as 2 h.

11:45 to 12:00 is 15 min.   12:00 to 14:00 is 2 h.   14:00 to 14:20 is 20 min.

      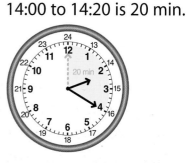

Total time: 15 min + 2 h + 20 min = 2 h 35 min

Practice

1. Which clocks show the same time?

a)

P.M.

b)

c)

d)

e)

P.M.

f)

A.M.

2. Write each time using A.M. or P.M.

a)

b)

c)

d)

3. Write each time using a 24-h clock.

a)

P.M.

b)

P.M.

c)

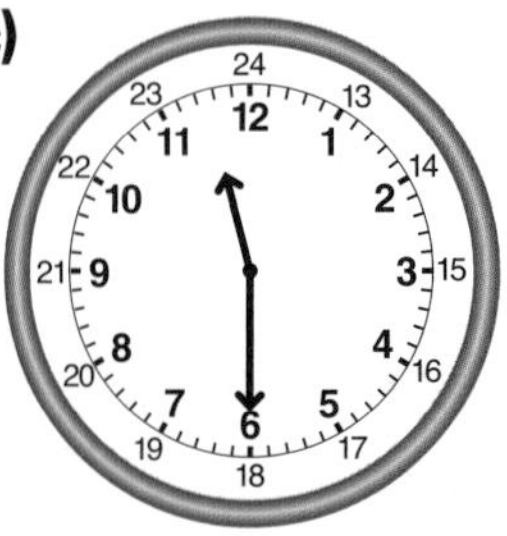

P.M.

d)

A.M.

4. Sandeep picked up his cousin at the airport.
Her flight arrived at 19:40.
What time is this on a 12-h clock?

5. a) A ferry leaves Port Hardy at 07:30 and arrives in Prince Rupert at 22:45.
How long is the journey?

b) An overnight ferry leaves Shearwater at 23:45 and arrives at
Port Hardy at 08:10. How long is the journey?

6. A bus leaves Anatown at 11:50 A.M. and
arrives in Beaconsfield 3 h 25 min later.
What time does the bus arrive in Beaconsfield?
Show the time as many different ways as you can.

7. A bus leaves Halifax, Nova Scotia, at 09:50.
It arrives in Sydney, Nova Scotia, at 16:20. How long is the trip?

8. Michel is flying from Montreal, Quebec, to Fort
Lauderdale, Florida. The flight leaves at 16:15.
Passengers to the United States should check in
at least 1 h 30 min before their flights leave.
Michel's watch shows the time he arrived at the airport.
Is he on time? Explain how you know.

Reflect

When you write or tell a time, which way do you prefer:
using a 12-h clock or a 24-h clock?
Give examples in your answer.

Covering Shapes

Louis counts how many
blue Pattern Blocks it takes
to cover this star.

Explore

You will need Pattern Blocks.
Estimate how many blue Pattern Blocks cover each shape.
Cover each shape to check your estimate.
Order the shapes by the number of blocks that cover them, from greatest to least.
Show your work.

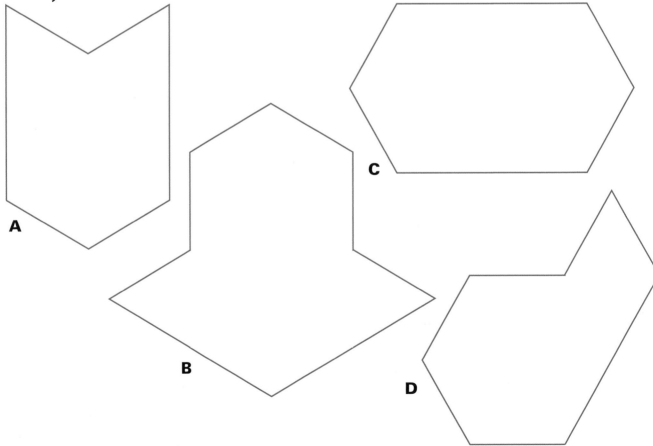

Show *and* Share

Share how you ordered the shapes.
How close were your estimates to the actual numbers
of blocks?

Connect ....

The number of units needed to cover a shape is the **area** of the shape.
The units must be the same size. The units must be *congruent*.
You can find the area of a shape by counting how many units cover it.

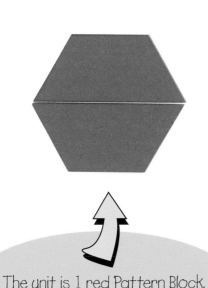

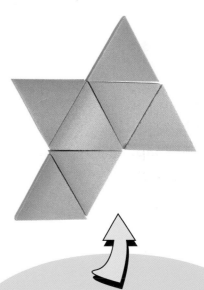

The unit is 1 red Pattern Block.
The area is 2 red Pattern Blocks.

The unit is 1 green Pattern Block.
The area is 7 green Pattern Blocks.

1. Make each shape with Pattern Blocks.
   Use 1 green Pattern Block as the unit.
   Find the area of each shape.

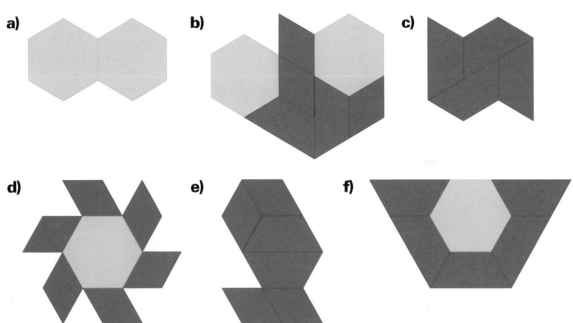

a) b) c)

d) e) f)

2. Make your own shape with Pattern Blocks.
   Draw the shape on triangular dot paper.
   Have a classmate find the area of the shape.
   She can choose the units.

3. Use red and yellow Pattern Blocks.
   a) Make a design with
      area 12 green Pattern Blocks.
   b) Make a design with area between
      10 and 15 green Pattern Blocks.
   c) Make a design with
      area 6 blue Pattern Blocks.
   d) Make a design with
      area 9 blue Pattern Blocks.
   Colour triangular grid paper to show your designs.

4. The area of a shape is 6 green Pattern Blocks.
Draw the shape on triangular dot paper.
How many different shapes can you make?
Explain how you made the different shapes.

5. Use Pattern Blocks to find the area of this fish.
The unit is 1 blue Pattern Block.

6. Suppose the unit for area is 1 green Pattern Block.
How can you find the area of the fish above without
using green Pattern Blocks? Explain.

7. Use Tangram pieces.

Find the area of each shape in small triangles.
a) the medium triangle
b) the square
c) the quadrilateral that is *not* the square
d) the large triangle

Reflect

Suppose you know how many blue Pattern Blocks cover
a shape. How can you find how many green Pattern
Blocks cover the same shape?
Use words, pictures, or numbers to explain.

Exploring Area

You will need Pattern Blocks and a small book.

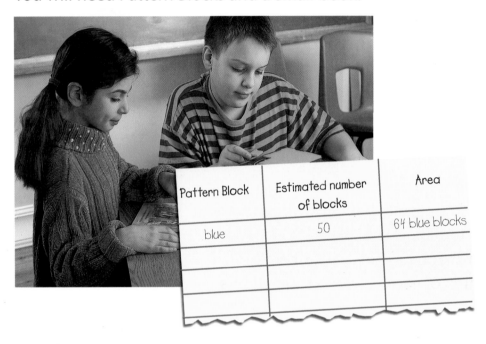

Pattern Block	Estimated number of blocks	Area
blue	50	64 blue blocks

➤ Estimate how many blue Pattern Blocks will cover the surface of the book.

➤ Cover the book. Find its area in blue Pattern Blocks.

➤ Record your work in a table.

Repeat this activity with:
• red Pattern Blocks
• orange Pattern Blocks
• green Pattern Blocks

Show *and* Share

Which Pattern Blocks worked best to cover the book? Why?
What did you do when the blocks did not cover the book completely?
Share your ideas with your partner.

Connect

The area of a surface is usually measured in **square units**.

➤ To find the area of a surface,
you can count the number of
square units.
The area of this patio
is 6 square units.

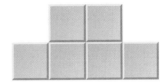

Squares fit together well. They are easy to count.

➤ To find the area of a rectangle,
you can count the square units or multiply.
There are 4 rows of 3 squares.
$4 \times 3 = 12$

The area of this rectangular patio
is 12 square units.

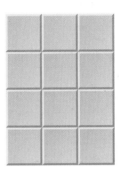

Practice

1. Estimate which shape has the greatest area.
Then find the area of each shape in square units.

a)    b)   c) 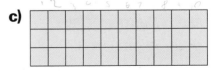   d)

2. Order the shapes in question 1 from least to greatest area.

3. Write a multiplication fact to find the area of each rectangle.

a)   b)   c)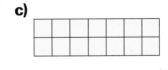

d) 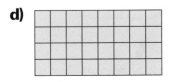   e)    f)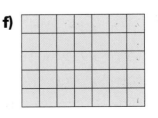

4. Find the area of each game board in square units.
Write a multiplication fact for each area.

a) Checkers

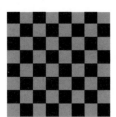

b) Snakes and Ladders

5. The area of a rectangle is 32 square units.
The rectangle has 8 rows of squares.
How many squares are in each row? How do you know?

6. Estimate which shape has the greatest area.
Then find each area in square units.
How can you do this by multiplying, then adding?

a)

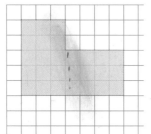

b)

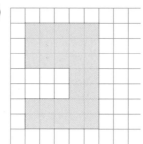

c)

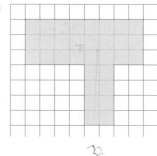

7. Use grid paper.
Draw a rectangle with each area.

a) 12 square units   **b)** 7 square units

c) 15 square units   **d)** 9 square units

8. Kelly drew a shape with an area of 48 square units.
What might the shape look like?
Use words, pictures, or numbers to show your ideas.

Reflect

Explain why the square unit is best for finding area.

Measuring Area in Square Centimetres

You will need cardboard rectangles and a transparent 1-cm grid.

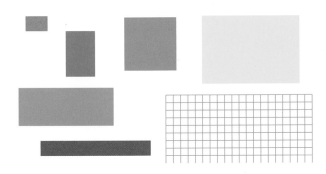

➤ Choose a rectangle. Estimate its area in centimetre squares.

➤ Place the transparent grid over the rectangle.
Make sure you line up the sides of the rectangle with the lines on the grid.

➤ Find the area of the rectangle.

➤ Record your work in a table.

➤ Repeat the activity with the other rectangles.

➤ Order the rectangles from least to greatest area.

Rectangle	Estimated area	Actual area
Yellow	20 squares	

Show *and* Share

How could you use the area of one rectangle to estimate the area of another?
Why must the grid lines line up with the sides of the rectangle?

LESSON FOCUS | Use a 1-cm grid to measure area in square centimetres.

153

Each side of every square on this grid paper is 1 cm long.

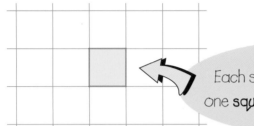

Each square has an area of
one **square centimetre** ($1\ cm^2$).

You can use square centimetres to measure area.
This rectangle is drawn on 1-cm grid paper.
It has 2 rows of 3 squares.
$2 \times 3 = 6$
The area of the rectangle is $6\ cm^2$.

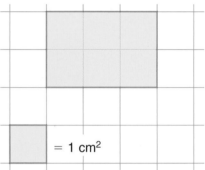

$= 1\ cm^2$

One face of a Base Ten unit cube has an area of $1\ cm^2$.
You can cover a shape with unit cubes,
then count the cubes to find the area in square centimetres.

My fingerprint has an area of about 1 square centimetre. I can use my fingerprint as a benchmark to estimate the area of this square. I think about how many of my fingerprints will fill the square.

Practice

1. Name 2 different benchmarks you could use to estimate area in square centimetres. Explain your choices.

2. Use one of your benchmarks.
   Estimate the area of each object in square centimetres.
   Then use a transparent 1-cm grid to find the approximate area.
   a) the top of a calculator   **b)** the cover of a small book

3. Roland drew this robot's head on 1-cm grid paper.
   a) What is the area of the head?
   b) What is the area of one robot eye?
      Its nose? Its mouth?

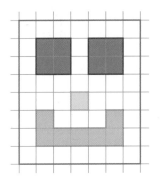

4. Draw your own shape on the lines of 1-cm grid paper.
   Find the area of your shape. Explain your strategy.

5. Use 1-cm grid paper. Draw a rectangle with each area.
   a) 8 cm^2 **b)** 24 cm^2 **c)** 16 cm^2 **d)** 18 cm^2

6. Look at this rectangle.
   Use 1-cm grid paper.
   a) Draw a rectangle with a greater area.
   b) Draw a rectangle with a lesser area.
   c) Double the length and width of the
      rectangle you drew for part b.
      Draw a new rectangle. Record its area.

7. You will need 1-cm grid paper, a number cube
   labelled 1 to 6, and 2 different colour crayons.
   Outline a 10 by 10 square on the grid paper.
   Take turns to roll the number cube.
   The number you get is the area in
   square centimetres you colour on the grid.
   For example, if you roll a 3, you colour in:

Game

or   but not

Continue to play until you have filled the 10 by 10 square.
Count the squares to find who coloured in the greater area.

Reflect

Draw a rectangle on 1-cm grid paper.
Find its area.
Explain your strategy.

Estimating and Measuring Area

You will need cardboard or plastic polygons and 1-cm grid paper.

Choose 3 polygons.

➤ Choose a benchmark for square centimetres.
   Use your benchmark to estimate the area of each polygon.

➤ Trace each polygon onto 1-cm grid paper.

➤ Count squares and parts of squares to find an approximate area.
   Record the approximate area of each tracing.

Show *and* Share

Share your strategies for counting part squares.
What do you do when part of a square is
greater than $\frac{1}{2}$ a square? Less than $\frac{1}{2}$ a square?

2 half squares make
1 whole square.

Math Link

Social Studies

Patchwork quilts were an early form of recycling.
They were made from leftover fabric and pieces cut
from old clothing.
How many squares are on this patchwork quilt?
How can you find out without counting every square?

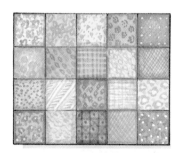

This triangle is drawn on 1-cm grid paper.

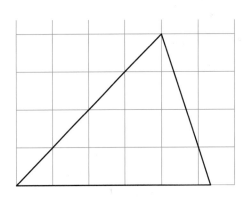

Here is one way to find the approximate area of this triangle.

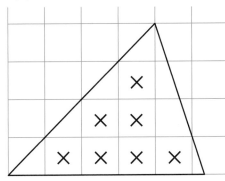

Count the whole squares.
There are 7 whole squares.

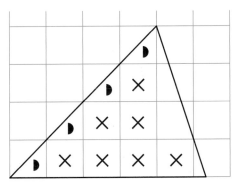

Put half squares together to count as whole squares.
There are 4 half squares.
4 half squares = 2 whole squares

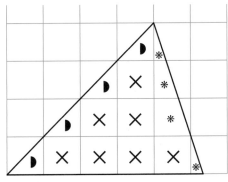

For parts of squares that are not half squares:
If the part is greater than $\frac{1}{2}$ a square, count it as 1 square.
If the part is less than $\frac{1}{2}$ a square, ignore it.
There are about 2 more squares.

Find the total number of squares:
$7 + 2 + 2 = 11$
The area of the triangle is about 11 cm^2.

1. Find the approximate area of each polygon.

   a) 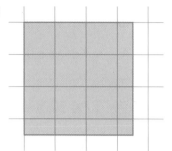   **b)** 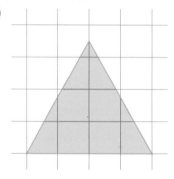   **c)**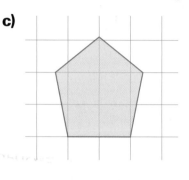

2. Order the polygons in question 1 from greatest to least area.

3. Use 1-cm grid paper. Draw a polygon with each area.
   a) about 10 cm²   **b)** about 12 cm²   **c)** about 19 cm²

4. Use 1-cm grid paper.
   Draw 3 different polygons with an area of about 15 cm².

5. Draw this face on 1-cm grid paper.
   Find the area of each part of the face.
   a) one eye
   b) the nose
   c) the mouth
   d) the whole face

   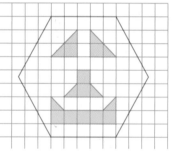

 6. Copy this polygon onto 1-cm grid paper.
   Explain how you would find the
   approximate area of this polygon.
   Show your work.

   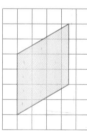

Reflect

You have estimated area and measured area.
When might an estimate be good enough?
When might you need to know the area precisely?
Write about your ideas.

Finding Area in Square Metres

Explore

You will need newspapers, tape, scissors, and a metre stick.

➤ Use the materials above to make a square with side length 1 m for each group member.

➤ Estimate the areas of different parts of your classroom or school. Then use your metre squares to find the areas.

➤ Record your results.

Show and Share

What did you do when the area was not an exact number of metre squares? Show how you can order the areas you measured from least to greatest.

Connect

Each square you made has an area of one **square metre**. You write one square metre as 1 **m²**.

You can use square metres to measure the area of a large surface, such as a soccer field.

This table top has an area of about one square metre. I can use it as a benchmark to estimate the area of a large surface. I think about how many of it will cover the surface.

You can use grid paper to model a large area.
On this grid, the area of 1 small square represents 1 m².

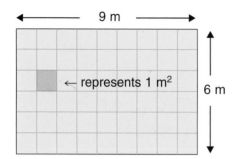

9 m

← represents 1 m²

6 m

This is a model of the floor of a gymnastics centre. It is 6 m wide and 9 m long.
The model has 6 rows of 9 squares.
$6 \times 9 = 54$

The area of the floor is 54 m².

Practice

1. Name a benchmark you could use to estimate area in square metres. Explain your choice.

2. Which unit – square centimetre or square metre – does each benchmark represent?
   a) a Smartie   **b)** a sidewalk square   **c)** a calculator key
   d) a table top   **e)** your front tooth   **f)** a small button

3. Work with other students to build a rectangle with each area.
   Use grid paper to draw a model of each rectangle.
   a) 12 m²   **b)** 9 m²   **c)** 14 m²

4. Find the area of each rectangle.

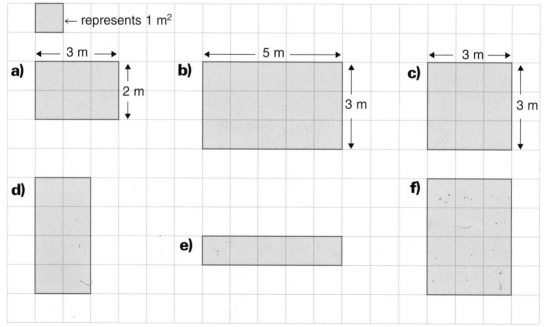

← represents 1 m²

a) 3 m, 2 m   **b)** 5 m, 3 m   **c)** 3 m, 3 m

d)   **e)**   **f)**

5. In question 4, which rectangle would need the most paint to cover it? The least paint to cover it?

6. The area of a rectangular garden is 24 m². The garden is 6 m long.
  a) How wide is the garden?
  b) Draw a model of the garden on 1-cm grid paper.

7. Which benchmark would you use to estimate the area of each item? Explain your choice.
  a) a classroom wall **b)** a page of your math book
  c) a photograph **d)** a flower garden
  e) a kitchen floor **f)** a table tennis table

8. Which measurement unit would you use to find the area of each item in question 7?

9. Here is a map of the playground at Peekaboo Day Care Centre.

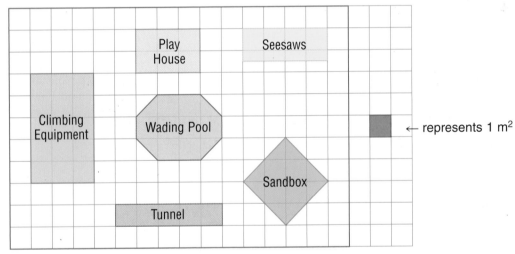

  a) Find the area of each section of the playground. Record your work in a table.
  b) Which section of the playground has the least area? The greatest area? How do you know?
  Show your work.

Reflect

How do you decide whether to use square centimetres or square metres to find the area of a surface?
Use numbers, pictures, or words to explain.

Strategies Toolkit

LESSON 12

Explore

Zoe bought 4 large squares of plywood to make the floor of a pen for her rabbits. She arranges the squares so that whole sides are touching. Find all possible shapes for the floor of the pen.

Show *and* Share

Explain how you solved the problem.

Connect

Brad has twelve 1-m squares of plywood to make the floor of a pen for his dog.
Brad wants to make a rectangular pen.
How many different rectangular floors can Brad make?
How much fencing would each pen need?

Strategies

- Make a table.
- Use a model.
- Draw a picture.
- Solve a simpler problem.
- Work backward.
- Guess and test.
- Make an organized list.
- Use a pattern.

Understand

What do you know?
- The pen will be made with 12 squares.
- The pen must be a rectangle.

Plan

Think of a strategy to help you solve the problem.
- You can **draw a picture**.
- Use grid paper.

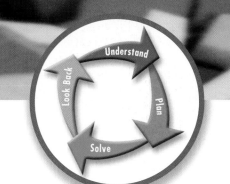

- How many different rectangles can you draw with 12 squares?
- How did you find the length of fencing for each pen?

- Did you find all the possible rectangles? How do you know?
- How could you solve the problem a different way?

Practice

Choose one of the **Strategies**

1. Suppose Zoe bought only 3 squares of plywood for the floor of her rabbit pen.
   a) How many different shapes could she make the pen?
   b) How much fencing would she need for each pen?

2. Calvin arranges 6 matching square concrete slabs to make a patio. Use grid paper.
   Draw all the possible patios Calvin could make.

3. Raji has 13 matching square concrete slabs. She wants to make a rectangular patio.
   a) Draw Raji's patio.
   b) Is there more than one possible patio? Explain.

Reflect

Use words, pictures, or numbers to explain how you can draw a picture to solve a problem.

Exploring Rectangles with Equal Areas

Explore

You will need 48 Colour Tiles or congruent squares, and grid paper.
Each tile or square represents 1 m^2.

Ms. Daisy is planning a rectangular garden for her backyard.
The garden will have an area of 48 m^2.

➤ Use the tiles or squares to find all the
possible rectangles that Ms. Daisy can make.
➤ Draw a model of each rectangle on
grid paper.

Show *and* Share

How many different rectangles did you find?
Tell what you know about the area of each rectangle.
Which rectangle would you recommend to Ms. Daisy?
Explain your choice.

Connect

Different rectangles can have equal areas.
Each rectangle below has area 12 m^2.

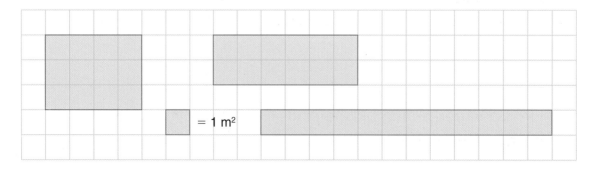

= 1 m^2

Use Colour Tiles or congruent squares when they help.

1. Use 1-cm grid paper. Draw all possible rectangles with each area.
   a) 1 cm²   b) 7 cm²   c) 24 cm²   d) 16 cm²
   e) 13 cm²   f) 6 cm²   g) 15 cm²   h) 20 cm²

2. Copy each rectangle onto 1-cm grid paper.
   • Find the area of each rectangle.
   • Draw a different rectangle with the same area.

a)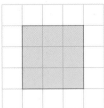

b)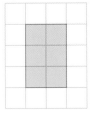

c)

d)

e)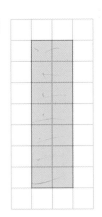

f)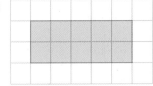

3. Mrs. Patel's rectangular patio has area 18 m².
   Draw a model on grid paper to show what you think
   Mrs. Patel's patio looks like. Explain your decision.

4. The area of a rectangular garden is 36 m².
   Draw models of all the possible rectangles for the garden.
   Have you drawn all the possible rectangles? How do you know?

Reflect

How can you use multiplication facts to sketch all possible
rectangles with area 40 cm²?

Unit 4 Show What You Know

1. Write each date in metric notation.
  a) November 23rd, 2008   **b)** February 8th, 1941   **c)** March 11th, 2007

2. Write each date using words and numbers.
  a) 1999 12 07   **b)** 2002 04 17   **c)** 1866 12 25

3. Rana wrote her birthday as 09 03 93.
  a) When might her birthday be?
  b) Rana said that her birthday was the first day of school.
     Write Rana's birthday in metric notation.

4. Write each time in two ways.

  a)    **b)**    **c)** 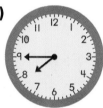   **d)**

5. Draw a digital clock to show each time.
  a) quarter past 3   **b)** half past 10
  c) two fifty-three   **d)** quarter to 4

6. Dennis played basketball for 40 minutes.
   He started to play at 3:15.
   Draw a digital clock to show what time he stopped.

7. The movie started at 4:20 p.m.
  a) Sami was 10 minutes late. What time did he arrive?
  b) Sofia was 7 minutes early. What time did she arrive?

8. Write each time using the 24-hour clock.
  a) 9:43 A.M.   **b)** 8:24 P.M.   **c)** 4:25 A.M.

9. Write each time using the 12-hour clock.
  a) 05:15   **b)** 14:20   **c)** 23:42

7

10. Use red and blue Pattern Blocks.
  a) Make a design with area 20 green Pattern Blocks.
  b) Make a design with area between 4 and 6 yellow Pattern Blocks.
  Colour triangular dot paper to show your designs.

8

11. The area of a rectangle is 18 square units.
  The rectangle has 3 rows of squares.
  How many squares are in each row?
  How do you know?

9

12. Use 1-cm grid paper.
  Draw a rectangle with each area.
  a) 21 cm^2 **b)** 25 cm^2 **c)** 30 cm^2

10

13. Find the approximate area of this polygon.
  Explain your strategy.

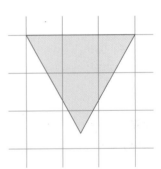

11

14. The area of a rectangular garden is 27 m^2.
  The garden is 9 m long.
  a) How wide is the garden?
  b) Draw a model of the garden
     on 1-cm grid paper.

15. Which benchmark would you use to
  estimate the area of each item?
  Explain your choice.
  a) a page of your favourite book
  b) an ice rink
  c) the school parking lot
  d) your hand print

UNIT

4 Learning Goals

☑ read and record time using
  analog and digital clocks

☑ read and record
  calendar dates

☑ estimate and measure area

☑ construct rectangles for a
  given area

13

16. Use 1-cm grid paper.
  Draw all the possible rectangles with
  each area.
  a) 2 cm^2 **b)** 14 cm^2 **c)** 8 cm^2

Design a Playground

The Little Owl Daycare Centre needs a new playground.
Design the playground.

Here are the guidelines:
- The playground has the shape of a rectangle.
- It has sections for 4 or 5 pieces of equipment.
- The sections are far enough apart to make sure the children are safe.
- You may include other features in your playground.

- ➤ Use grid paper to draw a plan for your playground.

- ➤ Let the area of one grid square represent 1 m^2.

- ➤ Make a table to go with your plan.
  The table should show the area of each section.

- ➤ Explain your plan.

Reflect on Your Learning

What have you learned about telling time?
How do you find the area of a shape?
Use words, pictures, or numbers to explain.

The Icing on the Cake

You will need Snap Cubes.

My piece has icing on three faces.

My piece has icing on one face.

My piece has icing on two faces.

Part 1

➤ Use Snap Cubes to model different-sized, square, one-layer "cakes."
Start with a 2 by 2 square.
Make larger and larger squares.

➤ Imagine that the top and sides are covered in icing.
Each Snap Cube represents one piece of cake.

➤ For each cake, how many pieces have icing on one face?
How many pieces have icing on two faces? Three faces?
Record your work in a table.

Size of Cake	Number of Pieces	Number of Faces with Icing		
		3 Faces	2 Faces	1 Face
2 by 2	4			

Part 2

➤ What patterns can you find in the table?

➤ What do you notice about the numbers of pieces that have icing on three faces?

➤ One cake has 16 pieces with icing on two faces. How many pieces are there in the whole cake?

➤ Suppose you made a 10 by 10 cake. How many of each kind of piece would there be?

Display Your Work

Make a poster display to show all the patterns you found.

Take It Further

Model rectangular cakes with 24 Snap Cubes.

➤ How many different cakes can you model using all 24 cubes?

➤ How many cakes have pieces with icing on four faces?

➤ Which cake has the greatest number of pieces with icing on one face? Two faces? Three faces?

➤ Write about what you found out.

Spring Activities Day

Learning Goals

- model, name, and record fractions of a whole and of a set
- compare and order fractions
- relate tenths and hundredths as decimals and fractions
- explore equivalent decimals
- use decimals to record money values
- add and subtract decimals and money
- identify everyday contexts in which fractions and decimals are used

Decimals

The Grade 4 students are holding an Activities Day to celebrate the arrival of spring.

In the Egg Race, the winner is the person who puts the most plastic eggs into the cartons. Here are some results.

Name	Carton Filled
Penny	$\frac{4}{12}$
Maria	$\frac{11}{12}$
Brady	$\frac{7}{12}$

- What does the fraction $\frac{4}{12}$ mean?
- Where have you seen fractions like those in the table? How were they used?
- What are some questions you can ask about the Egg Race?

Key Words

fifths

tenths

numerator

denominator

proper fraction

unit fraction

decimal

decimal point

one-hundredth

equivalent decimals

Fractions of a Whole

Pioneers made quilts from scraps of
material sewn together.
Square pieces are easy to fit together.

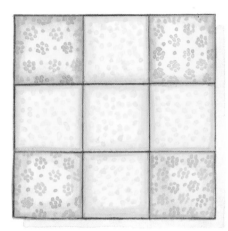

Explore

You will need Colour Tiles or congruent squares, and 1-cm grid paper.
Design a quilt that uses squares of at least 3 colours.
Record your quilt on grid paper.
Use fractions to describe your quilt.

Show **and** Share

Trade quilts with your partner.
Describe your partner's quilt using words and fractions.
How did you know which fractions to use?

Fractions name equal parts of a whole.

2 equal parts are halves.
$\frac{1}{2}$ or one-half is blue.

4 equal parts are fourths.
$\frac{2}{4}$ or two-fourths are green.

5 equal parts are **fifths**.
$\frac{5}{5}$ or five-fifths are yellow.

10 equal parts are **tenths**.
$\frac{4}{10}$ or four-tenths are red.

$\frac{4}{10}$ of the rectangle are red.
$\frac{6}{10}$ of the rectangle are white.

10 is the **denominator**.
It tells how many
equal parts are in 1 whole. ⟶

$$\frac{4}{10}$$

4 is the **numerator**.
It tells how many
equal parts are counted.

A fraction is a number.

$\frac{4}{10}$ is a **proper fraction**.
It represents an amount less than 1 whole.

$\frac{5}{5}$ is a fraction.
It represents an amount equal to 1 whole.

Practice . . . .

1. Use Colour Tiles or congruent squares to show each fraction.
   Record your work on grid paper.
   a) $\frac{2}{3}$ b) $\frac{6}{10}$ c) one-eighth   d) four-fifths

2. Write a fraction to tell what part of each quilt is striped.

a) 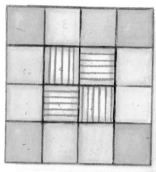 b) 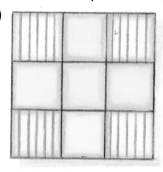 c)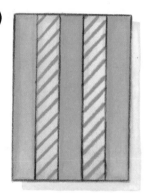

3. Write a fraction to tell what part of each quilt in question 2 is *not* striped.

4. Which pictures show thirds? How do you know?

a) 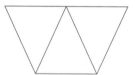 b)  c)

5. Does this diagram show $\frac{3}{4}$? Explain.

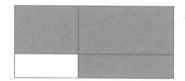

6. Write a fraction to describe the shaded part of each shape.

a)   b)   c)   d)

7. Write a fraction to tell what part of each shape in question 6 is *not* shaded.

8. About what fraction of each container is filled?

a)  b)  c)

9. Use a copy of these shapes.
Shade each shape to show the given fraction.

a) Show $\frac{1}{2}$.

b) Show $\frac{7}{8}$.

c) Show $\frac{3}{4}$.

d) Show $\frac{2}{3}$.

e) Show $\frac{5}{6}$.

f) Show $\frac{2}{2}$.

10. Look around your classroom.
Find an item that is divided into equal parts.
Sketch the item. Name the equal parts.

11. Use grid paper.
Design a place mat by colouring squares in a rectangle.
Use fractions to describe your place mat design.

 12. Nadia shares her square brownie with Tran.
Here are two ways to cut the brownie.
Are the pieces the same size?
Show your work.

Math Link

Physical Education

A volleyball court is divided into 2 equal parts. What other sports are played on a region that is divided into equal parts?

Reflect

Use pictures, words, or numbers to explain the fraction $\frac{5}{8}$.

Fraction Benchmarks

Explore

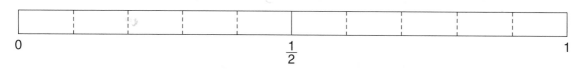

You will need 6 strips of squared paper the same length as this strip.

0                             $\frac{1}{2}$ 1

➤ Colour a paper strip to show $\frac{3}{10}$.
Line up your strip with the strip above.
Is $\frac{3}{10}$ closer to 0, $\frac{1}{2}$, or 1?

$\frac{3}{10}$	$\frac{9}{10}$	$\frac{6}{10}$
$\frac{1}{10}$	$\frac{7}{10}$	$\frac{2}{10}$

➤ Estimate if each fraction in the box is
closer to 0, $\frac{1}{2}$, or 1.
Use paper strips to check your estimates.
Record your findings in a table.

Closer to 0	Closer to $\frac{1}{2}$	Closer to 1

Show *and* Share

Talk with your partner about the fractions.
Which fraction is closest to 0? Closest to $\frac{1}{2}$? Closest to 1?

Connect

This number line shows the benchmarks 0, $\frac{1}{2}$, and 1.

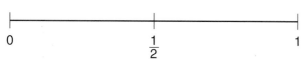

0           $\frac{1}{2}$ 1

You can use number lines with the same length to find
which benchmark each of $\frac{4}{5}$, $\frac{4}{10}$, and $\frac{4}{20}$ is closer to.

0          $\frac{1}{2}$ $\frac{4}{5}$ 1

 $\frac{4}{5}$ is closer to 1.

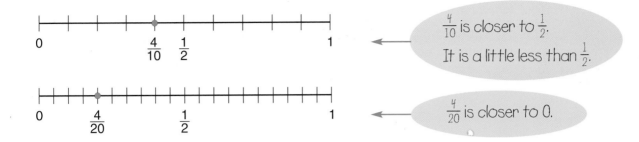

$\frac{4}{10}$ is closer to $\frac{1}{2}$.

It is a little less than $\frac{1}{2}$.

$\frac{4}{20}$ is closer to 0.

Practice

1. Is the fraction of juice left in each glass closer to 0, $\frac{1}{2}$, or 1?

   a)   b)   c)   d)

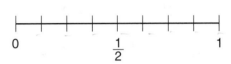

2. Use a copy of this number line.
   Place $\frac{7}{8}$, $\frac{1}{8}$, and $\frac{3}{8}$ on the number line.
   Which benchmark is each fraction closer to?

3. Name a fraction between $\frac{1}{2}$ and 1, but closer to 1.
   Draw a picture to show the fraction.
   Tell how you chose which fraction to draw.

4. Gary poured a little more than $\frac{1}{2}$ a glass of milk.
   Name a fraction that might tell how much milk that is.

5. Use a copy of this number line.
   Estimate to place a fraction:
   a) between 0 and $\frac{1}{2}$, but closer to 0
   b) between $\frac{1}{2}$ and 1, but closer to $\frac{1}{2}$

Reflect

Write two different fractions close to 0.
Use paper strips to show which fraction is closer to 0.

Exploring Fractions of a Set

Ten children are ready for art class.

What fraction of the group is girls?
What fraction of the group is wearing a striped shirt?
What other fractions can you use to describe the children?

Explore

You will need 2 colours of counters and a bag.

Put 10 counters of each colour in a bag.

➤ Take out a handful of counters.

➤ Use a chart.
Record fractions to describe the set of counters.

➤ Return the counters.

➤ Record your work.

Repeat the activity 5 times.

Number of counters pulled out	Fraction of counters that are red	Fraction of counters that are green

Show *and* Share

Talk with your partner.
Tell how you knew what fractions
to record for each set.

To find a fraction of a set, start by counting.

➤ There are 6 stars.
  5 of the 6 stars are yellow.
  $\frac{5}{6}$ of the stars are yellow.

➤ There are 12 spaces in the paint tray.
  8 of the 12 spaces have paint.
  $\frac{8}{12}$ of the spaces have paint.
  $\frac{4}{12}$ of the spaces are empty.

Practice •

Use counters when they help.

1. What fraction of each set is red?

   a)

   b)

   c)

   d)

2. Draw your own set of counters.
   Name the fractions in the set.

3. Copy each set of counters.
   Shade the counters to show each fraction.
   a) Show $\frac{7}{15}$.   b) Show $\frac{8}{8}$.   c) Show $\frac{1}{12}$.

4. Write a fraction to tell what part of each set in question 3 is *not* shaded.

5. What fraction of eggs are left in each carton?
   What fraction of eggs have been used?

a)  **b)**  **c)**

6. Joe has 5 rabbits.
   One-fifth is white.
   The rest is black.

   a) How many rabbits are white?
   b) How many rabbits are black?

7. Jordie has 3 pairs of black pants and
   2 pairs of red pants.
   What fraction of Jordie's pants are red?

8. Print your first name.
   a) What fraction of the letters in your first name are vowels?
   b) What fraction of the letters are consonants?

9. Louella had 2 black caps and 2 red caps.
   a) What fraction of her caps were red?
   b) Louella buys another cap.
     What fraction of her caps might be red now?
     Show your work.

10. Describe a situation in which you might use a fraction
   of a set.

Reflect

Explain why both pictures show $\frac{3}{10}$.

Finding a Fraction of a Set

Here is one way to arrange
12 counters to make equal groups.
How many other ways can you find?

3 equal groups of 4
Each group is $\frac{1}{3}$ of 12.

 **Explore**

You will need:
- 25 two-colour counters
- 8 fraction cards

You will use the counters to model
a fraction of a set.

➤ Shuffle the cards.
Put them face down in a pile.
➤ Take turns.
One person takes a fraction card.
Do not show the card.
➤ Count out the counters you need
to match your card.
Place all the counters red side up.
Turn some groups of counters yellow side up
to show your fraction.
➤ Have your partner tell you the fraction.
➤ Continue until you have used all the cards.

$\frac{3}{8}$ of 16

$\frac{4}{5}$ of 10

Show *and* Share

Tell the class about your strategy.
How did you know how many equal groups to make?

Fractions can show
equal parts of a set.

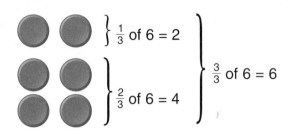

Here is a way to find $\frac{3}{5}$ of 10.

The denominator tells us
we are counting fifths.
Divide 10 counters into
5 equal groups to show fifths.

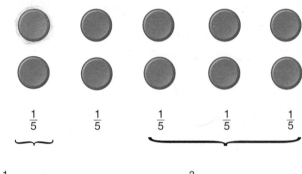

$\frac{1}{5}$ of 10 = 2 $\frac{3}{5}$ of 10 = 6

Practice

Use counters in questions 1 to 3.
Find the fraction of each set.

1. a) $\frac{1}{4}$ of 8   **b)** $\frac{2}{4}$ of 8   **c)** $\frac{3}{4}$ of 8

2. a) $\frac{1}{3}$ of 12   **b)** $\frac{2}{3}$ of 12   **c)** $\frac{3}{3}$ of 12

3. a) $\frac{2}{8}$ of 16   **b)** $\frac{4}{10}$ of 20   **c)** $\frac{3}{6}$ of 12

4. Draw a picture to show the fraction of each set.

   a) $\frac{2}{5}$ of 10   **b)** $\frac{3}{4}$ of 16   **c)** $\frac{5}{5}$ of 10

5. Find:

   a) $\frac{1}{2}$ of 10   **b)** $\frac{3}{4}$ of 12   **c)** $\frac{1}{5}$ of 5

6. Print the name of the town or region where you live.
   Use fractions to describe the letters in the name.

7. The pie shop sold 16 pies.
One-half of them were apple pies.
One-fourth of them were blueberry pies.
How many pies were not apple or blueberry?
Show your work.

8. 5 is $\frac{1}{4}$ of a set.
How many are in the set?

9. There are 10 boys in a class.
Two-fifths of the class are boys.
How many students are in the class?
How do you know?

10. When is $\frac{1}{2}$ of a set less than $\frac{1}{3}$ of another set?
When is it more?
Draw pictures to show your ideas.

11. When is $\frac{1}{4}$ of a group of children *not* equal to
$\frac{1}{4}$ of another group of children?
Use pictures, numbers, and words to explain your thinking.

Reflect

When might you want to find a fraction
of a set outside the classroom?

LESSON 5

Relating Fractional Parts of Different Wholes and Sets

Explore

You will need Cuisenaire rods or strips of coloured paper.

➤ Use the orange rod as one whole.
Find a rod that shows $\frac{1}{2}$ of the orange rod.
Draw a picture to record your work.

Find other pairs of rods so that
one rod is $\frac{1}{2}$ of the other rod.
Draw pictures to record your work.

Compare the rods that represent one-half.
Compare the rods that represent one whole.
What do you notice?

➤ Repeat the activity for:
  • pairs of rods that show $\frac{1}{3}$ of one whole.
  • pairs of rods that show $\frac{1}{4}$ of one whole.

Show and Share

Share your work with another pair of students.
Explain how you chose the rods. What did you find out about
the size of a fraction in relation to the size of the whole?

Connect

When 2 wholes have different sizes, the same fraction of the whole
is different for each whole.

➤ Show $\frac{2}{3}$ of each paper strip.

| $\frac{2}{3}$ | $\frac{1}{3}$ | $\frac{1}{3}$ | $\frac{1}{3}$ |

| $\frac{2}{3}$ | $\frac{1}{3}$ | $\frac{1}{3}$ | $\frac{1}{3}$ |

Two-thirds of the long strip is greater than $\frac{2}{3}$
of the short strip. The longer the whole, the
greater the length that represents $\frac{2}{3}$

➤ Show $\frac{3}{4}$ of 16 and $\frac{3}{4}$ of 12.

 $\frac{3}{4}$ $\frac{3}{4}$ of 16 counters are 12 counters.

 $\frac{3}{4}$ $\frac{3}{4}$ of 12 counters are 9 counters.

$\frac{3}{4}$ of 16 counters are greater than $\frac{3}{4}$ of 12 counters.
The greater the number of counters, the greater the
number that represents $\frac{3}{4}$

Practice

1. Use a 12-cm strip and a 6-cm strip.
   Fold and colour each strip to show $\frac{1}{6}$.
   Which strip shows a longer length that represents $\frac{1}{6}$?

2. Use a 9-cm strip and a 15-cm strip.
   Fold and colour each strip to show $\frac{2}{3}$.
   Which strip shows a longer length that represents $\frac{2}{3}$?

3. Draw a set of 20 apples and a set of 15 apples.
   Colour $\frac{2}{5}$ of each set of apples red.
   In which set does $\frac{2}{5}$ represent a greater amount? Explain.

4. Draw a picture to show:
   a) $\frac{2}{3}$ of one pie is greater than $\frac{2}{3}$ of another pie
   b) $\frac{3}{4}$ of one group of fish is less than $\frac{3}{4}$ of another group of fish

Reflect

What strategies help you compare the same fraction
of two different sets?

Strategies Toolkit

Explore

Nawar had 24 stickers.
He kept $\frac{1}{6}$ of the stickers
and divided the rest equally among his 5 friends.
How many stickers did each friend receive?

Show *and* Share

Describe how you solved this problem.

Connect

Kalpana had 40 prizes to award at the school fair.
She put $\frac{1}{4}$ of the prizes in the raffle.
She divided the rest equally among the 6 races.
How many prizes were there for each race?

Strategies

- Make a table.
- Use a model.
- Draw a picture.
- Solve a simpler problem.
- Work backward.
- Guess and test.
- Make an organized list.
- Use a pattern.

What do you know?
- There are 40 prizes.
- The raffle used $\frac{1}{4}$ of the 40 prizes.
- The rest of the prizes were for the winners of the 6 races.

Think of a strategy to help you solve the problem.
- You can **use a model**.
- Use counters to represent the prizes.

How many counters will you start with?
Remove the counters for the raffle.
How many counters are left?
Arrange them to show the prizes for the 6 races.
How many prizes are there for each race?

How could you solve this problem another way?
Try it, and use your answer to check your work.

Practice

Choose one of the **Strategies**

1. Gabriella bought 20 batteries.
   She put 2 batteries in her radio.
   Her 3 brothers divided the rest equally.
   How many did each brother receive?

2. Ms. Logan had gel pens to award after BINGO.
   She gave 4 to Tip, and said that he had $\frac{2}{6}$ of them.
   How many gel pens did Ms. Logan have to start?

3. Mabel coloured a rectangular sheet of paper $\frac{1}{8}$ purple,
   $\frac{1}{2}$ blue, and the rest yellow.
   What fraction of the paper is yellow?

Reflect

How can counters help you solve fraction problems?
Use words, pictures, or numbers to explain.

Comparing and Ordering Unit Fractions

Explore

Use Pattern Blocks to explore ways to compare fractions.

Use the yellow Pattern Block as 1 whole.
Which fraction in each pair is greater?

$\frac{1}{6}$ and $\frac{1}{1}$ $\frac{1}{2}$ and $\frac{1}{6}$

$\frac{1}{3}$ and $\frac{1}{2}$ $\frac{1}{6}$ and $\frac{1}{3}$

$\frac{1}{1}$ and $\frac{1}{3}$ $\frac{1}{2}$ and $\frac{1}{1}$

Show *and* Share

With your partner, talk about the ideas you used to choose your answers.
Record your work.

➤ A fraction with a numerator of 1 is a **unit fraction**.
$\frac{1}{2}, \frac{1}{10},$ and $\frac{1}{1}$ are unit fractions.

➤ With different unit fractions, the equal parts of the whole have different sizes.

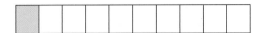   

$\frac{1}{10}$ ← 10 equal parts in the whole

$\frac{1}{4}$ ← 4 equal parts in the whole

Tenths are smaller than fourths.
So, $\frac{1}{10} < \frac{1}{4}$

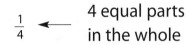

The more equal parts there are, the smaller the parts are.

➤ Order these unit fractions from least to greatest:

$$\frac{1}{8}, \frac{1}{3}, \frac{1}{4}$$

• One-eighth is the least because eighths are smaller
  than thirds and fourths.
• One-third is the greatest because thirds are greater than fourths.
• From least to greatest: $\frac{1}{8}, \frac{1}{4}, \frac{1}{3}$

You can draw pictures to check your thinking.

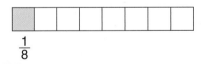

$\frac{1}{8}$

$\frac{1}{4}$

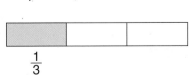

$\frac{1}{3}$

1. Look at each pair of shapes.
   Compare the fractional parts that are shaded.
   Write a fraction sentence using <, >, or =.
   a) 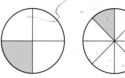   b)   c)

2. Draw pictures to show which is greater.
   a) $\frac{1}{2}$ or $\frac{1}{6}$ b) $\frac{1}{2}$ or $\frac{1}{4}$ c) $\frac{1}{3}$ or $\frac{1}{4}$
   d) $\frac{1}{1}$ or $\frac{1}{8}$ e) $\frac{1}{3}$ or $\frac{1}{2}$ f) $\frac{1}{10}$ or $\frac{1}{5}$

3. a) Which pair of students had the longest checkers game?
   b) Which had the shortest game?
      How do you know?

Checkers tournament	
Ivan vs. Fay	$\frac{1}{5}$ hour
Fay vs. Ari	$\frac{1}{2}$ hour
Fay vs. Cleo	$\frac{1}{3}$ hour

4. Use four 12-cm strips of paper.
   Show halves on one strip.
   Show sixths on one strip.
   Show quarters on one strip.
   Show thirds on one strip.
   Order these fractions from greatest to least: $\frac{1}{2}$, $\frac{1}{6}$, $\frac{1}{4}$, $\frac{1}{3}$
   Show your work.

5. Order these fractions from greatest to least.
   a) $\frac{1}{7}$, $\frac{1}{9}$, $\frac{1}{4}$ b) $\frac{1}{2}$, $\frac{1}{12}$, $\frac{1}{8}$ c) $\frac{1}{10}$, $\frac{1}{14}$, $\frac{1}{16}$ d) $\frac{1}{5}$, $\frac{1}{8}$, $\frac{1}{9}$

6. When can $\frac{1}{3}$ give you a bigger piece than $\frac{1}{2}$?
   Draw diagrams to explain.

At Home

Reflect

What strategies can you use to compare unit fractions?

Look in the kitchen, the bathroom, or the laundry area. Where do you and your family use fractions at home?

Comparing and Ordering Fractions with the Same Numerator or Denominator

Explore

You will need 4 paper strips each 20 cm long.

➤ Fold and colour strips to show each fraction.
$\frac{2}{4}$, $\frac{2}{8}$, $\frac{2}{3}$, $\frac{2}{6}$

$\frac{1}{4}$	$\frac{1}{4}$	$\frac{1}{4}$	$\frac{1}{4}$

$\frac{1}{3}$	$\frac{1}{3}$	$\frac{1}{3}$

➤ Compare the $\frac{2}{4}$ strip and the $\frac{2}{3}$ strip.
Which fraction is greater?
How do you know?

➤ Continue to compare
pairs of strips.
Record your findings
using > or <.

Show *and* Share

Talk with your partner
about the pairs of fractions.
Which of the 4 fractions is least?
Which is greatest?
How do you know?

LESSON FOCUS | Compare and order fractions with the same numerator or denominator.

193

➤ Different fractions of the same whole may have the same denominator. Then, the parts being counted have the same size.

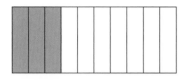 $\frac{3}{10}$ ⟵ Tenths are counted.

 $\frac{9}{10}$ ⟵ Tenths are counted.

The fewer the parts, the smaller the fraction
So, $\frac{3}{10} < \frac{9}{10}$

➤ Here is one way to order these fractions with the same denominator:
$\frac{3}{8}, \frac{7}{8}, \frac{2}{8}$

• $\frac{7}{8}$ is the greatest because it has the most parts.

• $\frac{2}{8}$ is the least because it has the fewest parts.

• From greatest to least: $\frac{7}{8}, \frac{3}{8}, \frac{2}{8}$

➤ Different fractions of the same whole may have the same numerator but different denominators. Then, the parts being counted have different sizes.

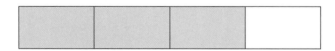 $\frac{3}{4}$ ⟵ Fourths are counted.

 $\frac{3}{6}$ ⟵ Sixths are counted.

The bigger the parts, the greater the fraction
So, $\frac{3}{4} > \frac{3}{6}$

➤ Here are 3 ways to compare and order fractions with the same numerator.

- Order $\frac{3}{5}$, $\frac{3}{10}$, and $\frac{3}{4}$ from greatest to least.
  Use strips of paper with the same length.
  Fold and colour to show each fraction.
  Line up the strips.

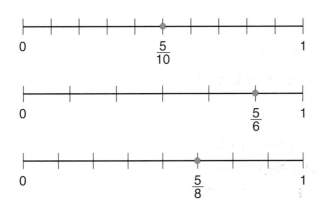

| $\frac{1}{4}$ | $\frac{1}{4}$ | $\frac{1}{4}$ | $\frac{1}{4}$ |

| $\frac{1}{5}$ | $\frac{1}{5}$ | $\frac{1}{5}$ | $\frac{1}{5}$ | $\frac{1}{5}$ |

| $\frac{1}{10}$ | $\frac{1}{10}$ | $\frac{1}{10}$ | $\frac{1}{10}$ | $\frac{1}{10}$ | $\frac{1}{10}$ | $\frac{1}{10}$ | $\frac{1}{10}$ | $\frac{1}{10}$ | $\frac{1}{10}$ |

The greatest fraction is the strip with the longest coloured part.
From greatest to least: $\frac{3}{4}$, $\frac{3}{5}$, $\frac{3}{10}$

- Order $\frac{5}{10}$, $\frac{5}{6}$, and $\frac{5}{8}$ from least to greatest.
  Draw number lines with the same length.

$\frac{5}{10}$ is to the left of $\frac{5}{8}$.

$\frac{5}{8}$ is to the left of $\frac{5}{6}$.
From least to greatest: $\frac{5}{10}$, $\frac{5}{8}$, $\frac{5}{6}$

- Compare $\frac{3}{4}$ and $\frac{3}{8}$.
  Use number lines and the benchmarks 0, $\frac{1}{2}$, and 1.

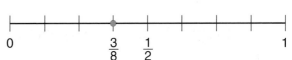

$\frac{3}{8}$ is between 0 and $\frac{1}{2}$.

$\frac{3}{4}$ is between $\frac{1}{2}$ and 1.

So, $\frac{3}{8} < \frac{3}{4}$

1. Use two 15-cm strips.
   Fold and colour one strip to show $\frac{2}{3}$.
   Fold and colour the other strip to show $\frac{2}{5}$.
   Which fraction is greater, $\frac{2}{3}$ or $\frac{2}{5}$?

2. Colour and fold two 16-cm strips to show $\frac{3}{4}$ and $\frac{3}{8}$.
   Use $>$ or $<$ to compare the fractions.

3. Draw a 12-cm number line.
   Label the number line with the benchmarks 0, $\frac{1}{2}$, and 1.
   Estimate to place $\frac{2}{3}$ and $\frac{2}{6}$ on the number line.
   Which fraction is greater? How do you know?

4. Use three 24-cm strips.
   Fold and colour the strips to show $\frac{3}{4}$, $\frac{3}{8}$, and $\frac{3}{6}$.
   Order the fractions from greatest to least.

5. Use a copy of these 3 number lines.
   Order $\frac{6}{10}$, $\frac{6}{8}$, and $\frac{6}{9}$ from least to greatest.

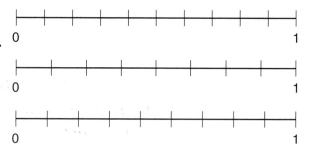

6. Olivia and Sayed each bought the same fruit bar.
   Olivia ate $\frac{2}{3}$ of her bar.
   Sayed ate $\frac{2}{4}$ of his bar.
   Use pictures, numbers, or words to show who ate more.

7. Jordan and Laci each has 12 coloured pencils.
   Jordan sharpened $\frac{3}{4}$ of his pencils.
   Laci sharpened $\frac{3}{6}$ of her pencils.
   Who sharpened more pencils? How do you know?
   Show your work.

Reflect

Which strategy do you prefer to compare fractions? Why?

Exploring Tenths

You have used Base Ten Blocks to model whole numbers.

What if you could use Base Ten Blocks to model fractions?

Explore

You will need Base Ten Blocks and grid paper.
Suppose a flat represents 1 whole.
What does a rod represent?

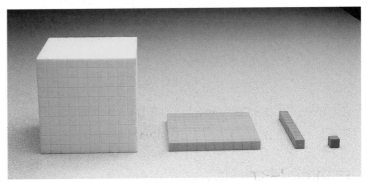

= 1

= ?

Use Base Ten Blocks to show these numbers.

$$\frac{6}{10} \qquad \frac{9}{10} \qquad \frac{3}{10}$$

Record your work on grid paper.

Show and Share

Share your work with another student.
Did you draw the same pictures for each number?
If you did not, who is correct?
Can both of you be correct? Explain.
What do you think a rod represents?

Here is one way to model $\frac{7}{10}$.

Any number in tenths can be written as a fraction or a **decimal**.
To write a fraction as a decimal, use a **decimal point**.

$\frac{7}{10}$ is the same as   0.7

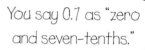
You say 0.7 as "zero and seven-tenths."

This is the decimal point.
Since $\frac{7}{10}$, or 0.7, are less than 1 whole,
we write 0 before the decimal point
to show there is no whole number part.

You can use a place-value chart to show a decimal.

Ones		Tenths
0		7

The decimal point is between
the ones place and the tenths place.

Practice

1. Write a fraction and a decimal for the coloured
   part of each picture.

a)

b)

c)

d)

e)

f)

2. Write a fraction and a decimal for the coloured part of each picture.

a)  **b)** 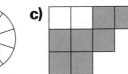 **c)** **d)**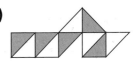

3. Draw a picture for each fraction.
Say the fraction, then write it as a decimal.

a) $\frac{4}{10}$ **b)** $\frac{7}{10}$ **c)** $\frac{9}{10}$

4. Draw a picture for each decimal. Write the decimal as a fraction.

a) 0.1 **b)** 0.8 **c)** 0.2

5. Write each number as a fraction and as a decimal.

a) nine-tenths **b)** three-tenths **c)** one-tenth

6. Alicia said 0.7 of the circle are red.
Is she correct? Explain how you know.

7. Write a decimal and a fraction for each letter on the number line.

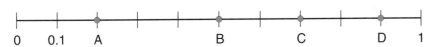

8. Look at the circle. Is each statement true or false? How do you know?

a) Two-tenths of the circle are red.

b) 0.3 of the circle are blue.

c) $\frac{6}{10}$ of the circle are not yellow.

d) 0.8 of the circle are blue or red.

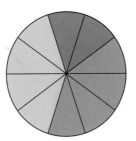

Reflect

Why is the decimal point important?
Use words, pictures, or numbers to explain.

Exploring Hundredths

Explore

You will need Base Ten Blocks and grid paper.

Suppose a flat represents 1 whole,
and a rod represents $\frac{1}{10}$.

What does a unit cube represent?

Use Base Ten Blocks to show these numbers.

$\frac{10}{100}$ $\frac{75}{100}$ $\frac{3}{10}$ $\frac{21}{100}$ $\frac{6}{100}$

Record your work on grid paper.

Show *and* Share

Share your work with another pair of students.
How did you find out what
a unit cube represents?

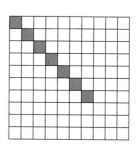

Connect

➤ This grid is divided into 100 equal squares.
Each square is **one-hundredth** of the grid.
Seven-hundredths of the grid are red.

We can write seven-hundredths as a fraction
or as a decimal.
$\frac{7}{100}$ is the same as 0.07

This zero shows there are   This zero shows there are
no whole number parts.   no tenths.

Ninety-one hundredths of this grid are yellow.
We write: $\frac{91}{100}$, or 0.91

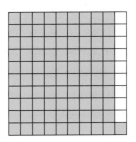

➤ We can use a place-value chart to show
a decimal with hundredths.

We say: zero and nine-hundredths
We say: zero and sixty-hundredths

Ones	Tenths	Hundredths
0	0	9
0	6	0

➤ We can use decimals to write parts of one dollar.
1 dollar = 100 cents
So, 1 cent = $\frac{1}{100}$ dollar, or 0.01 dollar

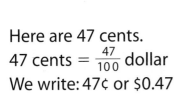

Here are 47 cents.
47 cents = $\frac{47}{100}$ dollar
We write: 47¢ or $0.47

We read and say 47¢
and $0.47 the same way:
forty-seven cents

Practice

1. Write a fraction and a decimal for the coloured part of each picture.

a)   b) 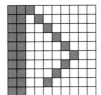   c)   d)

2. Colour a hundredths grid to show each number.
   a) $\frac{2}{100}$ b) $\frac{35}{100}$ c) 0.05   d) 0.18

3. Say each decimal, then model it with Base Ten Blocks.
   Record your work on grid paper.
   Label each picture with a fraction.
   a) 0.40   **b)** 0.31   **c)** 0.09   **d)** 0.02

4. Say each fraction, then write it as a decimal.
   a) $\frac{17}{100}$ **b)** $\frac{60}{100}$ **c)** $\frac{39}{100}$ **d)** $\frac{7}{100}$

5. Write as a decimal.
   a) six-hundredths   **b)** thirty-nine hundredths
   c) five-hundredths   **d)** fourteen-hundredths

6. Say each decimal, then write it as a fraction.
   a) 0.03   **b)** 0.16   **c)** 0.10   **d)** 0.54

 7. Write a fraction and a decimal
   for the coloured part of each grid.
   What do you notice about
   the coloured parts? Explain.
   Show your work.

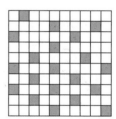 

8. Model each amount with dimes and pennies.
   Draw pictures to show your work.
   a) $0.46   **b)** $0.08   **c)** $0.21   **d)** $0.03

9. Write each amount as a decimal.
   a) 58¢   **b)** 9 cents   **c)** 73 cents   **d)** 14¢

10. Write each amount in words.
    a) $0.27   **b)** $0.18   **c)** $0.70   **d)** $0.01

11. Explain the meaning of each digit in each decimal.
    a) 0.11   **b)** 0.77   **c)** 0.44   **d)** $0.22

12. Describe a situation in which you might use hundredths in everyday life.

Reflect

Why are the zeros important in the decimals 0.5 and 0.05?

Equivalent Decimals

Explore

You will need Base Ten Blocks and hundredths grids.
Model each pair of decimals in as many ways as you can.

0.3 and 0.30   0.6 and 0.60

0.8 and 0.80   0.5 and 0.50

Record your work by colouring hundredths grids.

Show *and* Share

Share your work with another pair of students.
Discuss what you discovered about the pairs of decimals.

Connect

One row of this hundredths grid is one-tenth of the grid.
Each small square is one-hundredth of the grid.

= 1

4 rows are 4 tenths.

0.4

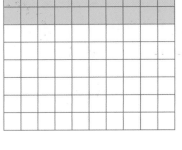

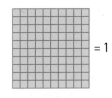

40 squares are
40 hundredths.

0.40

Both 0.4 and 0.40 name the shaded part of the grid.
So, 0.4 = 0.40
Decimals that name the same amount are called
equivalent decimals.

1. Write two equivalent decimals that name the shaded part of each grid.

a) 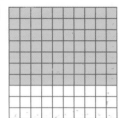 b) 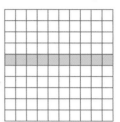 c) 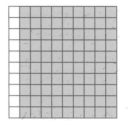 d)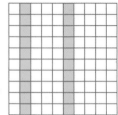

2. Colour hundredths grids to show each number.
   Write an equivalent decimal.
   a) 0.20   b) 0.9   c) 0.70   d) 0.1

3. Write an equivalent decimal for each number.
   a) 0.5   b) 0.80   c) 0.30   d) 0.6   e) 0.4
   f) 0.70   g) 0.90   h) 0.2   i) 0.50   j) 0.10

4. Find the equivalent decimals in each group.
   a) 0.5   0.05   0.50   b) 0.70   0.7   0.07
   c) 0.8   0.08   0.80   d) 0.04   0.40   0.4

5. Which bag of onions is the better buy?
   Explain how you know.

6. A student said that 0.40 is greater
   than 0.4 because 40 is greater than 4.
   Was the student correct? Use words,
   pictures, or numbers to explain.

7. Use dimes and pennies to show how many pennies are equal to 3 dimes.
   Draw a picture to show your thinking.

Reflect

When would you use tenths and hundredths
outside the classroom?

Adding Decimals to Tenths

The Base Ten flat represents 1 whole.
The rod represents one-tenth, or 0.1.

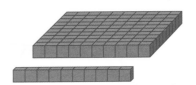

These blocks represent two and three-tenths.
We write: 2.3
We say: two and three-tenths

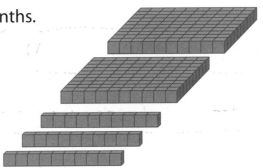

Explore

Ty enjoys hiking.
He keeps track of his weekend
hiking distances.
Estimate how far Ty walked
each weekend.
Find how far Ty walked each weekend.
Record your work.

Hiking Log!

WEEKEND	SATURDAY	SUNDAY
1	3.9 km	4.4 km
2	4.9 km	3.3 km

Show *and* Share

Show how you solved the problem.
What strategies helped you?
Share your ideas to start a class list.

Another weekend, Ty hiked 2.7 km and 1.8 km.

To estimate Ty's total distance, find a whole number close to each decimal.
2.7 is close to 3.
1.8 is close to 2.
3 + 2 = 5
Ty hiked approximately 5 km.

To add 2.7 + 1.8, use strategies similar to whole-number strategies.

➤ Use Base Ten Blocks to add 2.7 + 1.8.

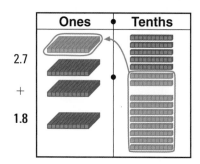

Add tenths and record.
Think:
15 tenths
equals
1 whole and
5 tenths

$$\begin{array}{r} 1 \\ 2.7 \\ +\ 1.8 \\ \hline 5 \end{array}$$

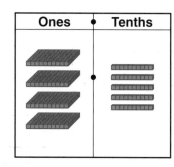

Add ones and record.

$$\begin{array}{r} 1 \\ 2.7 \\ +\ 1.8 \\ \hline 4.5 \end{array}$$

➤ Add from left to right.

2.7 + 1.8

$$\begin{array}{r} 2.7 \\ +\ 1.8 \end{array}$$

Add the ones: 2 + 1 = 3 ——————⟶ 3

Add the tenths: 0.7 + 0.8 = 1.5 ——⟶ 1.5

Add the sums: 3 + 1.5 = 4.5 ———⟶ 4.5

➤ Move by tenths to make one.

$$\begin{aligned} 2.7 + 1.8 &= 2.7 + 0.3 + 1.5 \\ &= \quad 3.0 \quad + 1.5 \\ &= 4.5 \end{aligned}$$

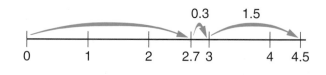

Ty hiked 4.5 km.

1. Estimate each sum.

   a) 0.4 + 0.3   **b)** 1.4 + 0.4   **c)** 2.6 + 1.1   **d)** 3.2 + 2.9

2. Add. Use Base Ten Blocks to help you.

   a) 1.8 + 2.1   **b)** 0.7 + 4.6   **c)** 3.6 + 1.2   **d)** 4.7 + 1.9

3. Will each sum be greater or less than 3? How do you know?

   a) 2.1 + 0.4   **b)** 2.3 + 0.9   **c)** 1.3 + 1.6   **d)** 1.2 + 2.1

Use the map below for questions 4 to 6.

4. Use the map to find the shortest distances.

	From:	To:
a)	Miller's Landing	Jake's Point
b)	Elma	Pearl
c)	Greenville	Elma
d)	Jake's Point	Port Baker
e)	Port Baker	Pearl

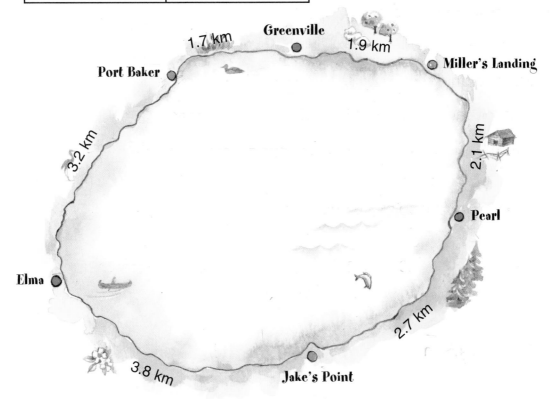

5. Franca travelled 6.8 km from one town to another town.
Where might Franca have travelled?
Show your work.

6. Make up a story problem that uses the map.
Solve your problem.

7. At 7:00 A.M., the temperature in Gander was 12.5°C.
By noon, the temperature had risen by 4.9°C.
What was the temperature at noon?

8. The Hon family buys fruit at the market.
Last Saturday, the family bought 2.6 kg of apples and
1.8 kg of bananas.
How much fruit did the family buy in total?

9. Alex's laptop computer has a mass of 2.3 kg.
The mass of the carrying case is 0.8 kg.
What is the total mass of the laptop and carrying case?

10. Marie-Claire rode her scooter
1.5 km to the store.
On the way home,
she took a different route that
was 0.7 km longer.
What was the total distance
Marie-Claire rode?
How do you know?

At Home

Reflect

Explain how adding decimals
is like adding whole numbers.
How is it different?

Look at the nutrition
information on a cereal box.
How are decimals used?
Write about what you notice.

Subtracting Decimals to Tenths

Explore

Liak is a long distance swimmer.
Her coach keeps track of her progress.

Liak's Progress Chart		
Class	Tuesday	Thursday
Week 1	1.4 km	2.8 km
Week 2	3.9 km	5.7 km

For each week, estimate how
much farther Liak swam
on the second day.
Then, find how much farther Liak swam.

Record your work.

Show and Share

Show how you solved the problem.
What strategies helped you?
Share your ideas to start a class list.

Suppose Liak swam 4.4 km on Thursday and 1.6 km on Tuesday.
How much farther did Liak swim on Thursday?

To estimate the distance:
4.4 is close to 4.
1.6 is close to 2.
$4 - 2 = 2$
Liak swam about 2 km farther on Thursday.

To subtract $4.4 - 1.6$, use whole number strategies.

➤ Use Base Ten Blocks
   to subtract.
   Trade 1 whole for
   10 tenths.

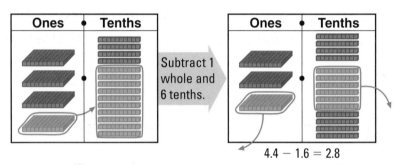

$4.4 - 1.6 = 2.8$

➤ Subtract from right to left.
   Try to subtract the tenths.

You cannot take 6 tenths from 4 tenths.	Trade 1 whole for 10 tenths.	Subtract the tenths.	Subtract the ones.
4.**4** − 1.**6**	³ ¹⁴ 4̸.4̸ − 1.6	³ **14** 4̸.4̸ − 1.**6** **.8**	³ **14** 4̸.4̸ − 1.**6** **2.8**

➤ Use mental math. Think addition.
   $1.6 + 0.4 = 2.0$
   $2.0 + 2.4 = 4.4$
   So, $4.4 - 1.6 = 0.4 + 2.4$
   $4.4 - 1.6 = 2.8$

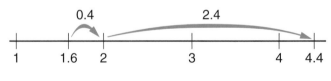

Liak swam 2.8 km farther on Thursday.

1. Subtract. You choose the strategy.
   a) 5.3 − 2.1   **b)** 4.9 − 0.7   **c)** 3.6 − 1.9   **d)** 7.4 − 4.8

2. Subtract.
   a) 52.7   **b)** 31.0   **c)** 9.6   **d)** 25.1
       − 45.8   − 5.7   − 2.7   − 12.2

3. Grant had 6.5 m of ribbon. He used 4.9 m to wrap up a gift.
   How much ribbon does Grant have left?
   How do you know your answer is reasonable?

4. Giorgio grew a 3.4-kg pumpkin. Toni grew a 4.1-kg pumpkin.
   Whose pumpkin had the greater mass? How much greater was it?

5. The temperature at 9:00 P.M. was 15.4°C.
   At midnight, it was 12.5°C.
   What was the change in temperature?

6. Aimee adopted a puppy from the Humane Society.
   Its mass was 4.7 kg. At the first visit to the vet,
   the puppy had a mass of 5.4 kg.
   How much mass had the puppy gained?

7. Jan ran the 100-m dash in 14.8 s.
   The school record is 15.6 s.
   By how much did Jan beat the school record?

8. Use estimation.
   Is the difference between 1.8 and 0.5 greater than 1
   or less than 1?
   Show your work.

Reflect

You know several ways to subtract two decimals.
Which way do you prefer?
Use words, pictures, or numbers to explain.

Adding and Subtracting Decimals to Hundredths

Money uses base ten with tenths and hundredths.

$1.00
or $1

$0.10
or 10¢

$0.01
or 1¢

Place value can help you read an amount of money.

Dollars (Ones)	Dimes (Tenths)	Pennies (Hundredths)

$2.54 or two dollars and fifty-four cents

Explore

Suppose you have $10.00 to spend.
What could you buy?
How much money would be left?

Show and Share

Compare your answers with a
classmate's answers.
How did you decide what to buy?

Fun Fair snack menu

☆ Hot Dog — $1.25
☆ Hamburger — $2.25
☆ Onion Rings — $2.25
☆ Pizza Slice — $0.99
☆ Yogurt — $0.99
☆ Fruit Cup — $1.25
☆ Water — $0.49
☆ Juice — $0.85
☆ Milk — $0.75

Lakshi had $5.
She bought a fruit cup for $1.25 and water for $0.49.

➤ How much did Lakshi spend?
Use a place-value mat to add $1.25 + $0.49.

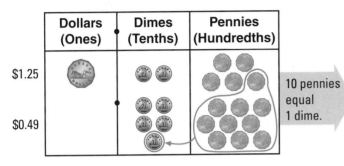

$1.25 + $0.49 = $1.74

Lakshi spent $1.74.

➤ What is Lakshi's change from $5?
• You can count on.

$1.74,   ... $1.75,   ... $2.00,   ... $3.00   ... $5.00
The change is $3.26.

• You can subtract from right to left.

Line up the decimal points.	Trade $1 for 10 dimes. Trade 1 dime for 10 pennies.	Subtract the cents.	Subtract the dollars.
$5.00 − 1.74	9 4 ̶1̶0̶ ̶1̶0̶ $5̶.0̶0̶ − 1.74	4 9 10 $5̶.0̶0̶ − 1.**74** **.26**	4 9 10 $5̶.0̶0̶ − 1.**74** **3.**26

The change from a $5 bill is $3.26.

1. Find each sum. You choose the strategy.
  a) $2.86 + $3.61   **b)** $4.79 + $3.18
  c) $6.27 + $3.04   **d)** $7.29 + $0.49

2. Add.
  a) $5.31   **b)** $4.11   **c)** $6.23   **d)** $5.60
        + 2.03   + 1.66   + 5.34   + 4.78

3. Find each difference. You choose the strategy.
  a) $5.75 − $2.51   **b)** $7.37 − $4.29
  c) $8.25 − $5.62   **d)** $3.25 − $0.18

4. What is the change from $10 when you spend each amount?
  a) $5.42   **b)** $3.76   **c)** $8.22   **d)** $4.20

5. Look at the items below. Find the cost of each pair of items.
  a)

  b)

  c)

  d)

$5.95 $7.95 $2.79 $1.89 $4.75

6. Look at the items above.
  a) About how much will all 5 beach items cost?
  b) About how much change will you get from $25.00?
  c) Find the exact cost and your change.
      How close were your estimates?
  Show your work.

7. Use the Ice Cold Drinks menu.
   a) Ira bought a milkshake. How much change did he get from $5.00?
   b) Suppose Ira bought water instead of a milkshake.
      How much money would he save?
   c) Jerry used a $5 bill to pay for one drink.
      His change was $2.21.
      What drink did Jerry buy?
      How do you know?

ICE COLD DRINKS

JUICE —————— $ 1.29
WATER —————— $ 1.10
MILKSHAKE ———— $ 2.95
ICE CREAM FLOAT — $ 2.79

8. Mei saved her allowance to go to the mall.
   She had $11.45 to spend.
   After two hours, Mei had spent $8.76.
   How much money did Mei have left?

9. Milo wants to buy some muffins.
   The cost is $5.95 plus tax. The tax is $0.36.
   Milo has $6.30.
   Does he have enough to buy the muffins?
   How do you know?

10. Hugh is a cashier.
    His cash register is out of pennies.
    Here is a customer's receipt.
    The customer pays with a $20 bill.
    Can Hugh make the correct change?
    Show your work.

```
THE GROCERY STORE

potatoes    $2.87
bread       $1.14
butter      $2.99
```

Reflect

You have learned two methods for making change.
When might you use each method?

LESSON

1

1. Write a fraction to tell what part of each figure is shaded.

a)   b)   c)

2. Write a fraction to tell what part of each figure in question 1 is *not* shaded.

2

3. Name a fraction between 0 and $\frac{1}{2}$, but closer to 0.
Draw a picture to show the fraction.

3

4. Tell what fraction of each set is shaded.

a)   b)   c)

4

5. Draw a picture to show the fraction of each set.
a) $\frac{2}{10}$ of 20 stars b) $\frac{4}{4}$ of 8 eggs c) $\frac{3}{5}$ of 10 squares

5

6. Draw a picture to show that $\frac{3}{4}$ of one pizza is *not* equal to $\frac{3}{4}$ of another pizza.

7. Draw a picture to show when $\frac{1}{3}$ of a group of fish is *not* equal to $\frac{1}{3}$ of another group of fish.

7

8. Draw a number line 12 cm long.
Label the benchmarks 0, $\frac{1}{2}$, and 1.
Estimate to place $\frac{1}{6}$, $\frac{1}{3}$, and $\frac{1}{10}$ on the number line.
Order $\frac{1}{6}$, $\frac{1}{3}$, and $\frac{1}{10}$ from greatest to least.

8

9. Order these fractions from least to greatest.
Use any materials to help you.

$\frac{3}{4}$　　$\frac{3}{8}$　　$\frac{3}{5}$　　$\frac{3}{10}$

9
10

10. Colour a hundredths grid to show each decimal.
a) 0.45 b) 0.09 c) 0.80 d) 0.3

11. Write an equivalent decimal for each decimal.

a) 0.8　　　**b)** 0.20　　　**c)** 0.1　　　**d)** 0.60

12. Kim had 2.6 m of blue fabric and 1.6 m of yellow fabric.

a) How much fabric did Kim have altogether?

b) How much more blue fabric did Kim have?

13. Add or subtract. Use Base Ten Blocks to help you.

a) 4.6 + 4.3　　　**b)** 3.4 − 1.2　　　**c)** 2.8 + 3.9

d) 1.7 − 1.5　　　**e)** 5.1 + 0.9　　　**f)** 1.3 − 0.8

14. Imagine you have $10.00 to buy school supplies.

$4.95　　$1.57　　$3.79

$5.25　　$2.68

a) Choose 2 items you want to buy. About how much will they cost? About how much money would you have left?

b) Which two items could you *not* buy with $10.00? Explain.

UNIT

5 Learning Goals

☑ model, name, and record fractions of a whole and of a set

☑ compare and order fractions

☑ relate tenths and hundredths as decimals and fractions

☑ explore equivalent decimals

☑ use decimals to record money values

☑ add and subtract decimals and money

☑ identify everyday contexts in which fractions and decimals are used

Spring Activities Day

You be the judge!
Here are the results for the top 3 students in each activity.

In the Egg Race, students had 3 minutes to carry the eggs, and fill the cartons. The winner was the person who filled the egg carton with the most eggs.

Egg Race

Name	Fraction of Carton Filled
Zachary	$\frac{5}{12}$
Wilma	$\frac{10}{12}$
Myles	$\frac{6}{12}$

In the Corn Cob Toss, students used an underhand throw to toss the corn cob as far as they could. The winner was the person with the longest toss.

Corn Cob Toss

Name	Distance
Percy–1st	4.99 m
Misty–2nd	4.68 m
Joi–3rd	4.45 m

In the Duck Waddle, students walked like a duck around the playground. The fastest person was the winner.

Duck Waddle

Name	Time
Maria–1st	29.8 s
Hunter–2nd	36.2 s
Thomas–3rd	45.3 s

Part 1

- Who won the Egg Race?
  Who came second? Third?
  How do you know?
- The results for the Corn Cob Toss
  were very close!
  What was the difference in the distances
  for 1st place and 2nd place?
  2nd place and 3rd place?
  How do you know?

Part 2

Make up a story problem about
the Spring Activities Day results.
Solve your problem.

Part 3

What event would you plan for a Spring
Activities Day?
How would you award the prizes?
Make up some examples to show
what might happen.
Use fractions or decimals in your activity.
Explain your work.

Check List

Your work should show
- ☑ how you used fractions
  and decimals correctly to
  find your answers
- ☑ your answers clearly
  explained
- ☑ how you developed a new
  successful event
- ☑ mathematical language
  and symbols used
  correctly

Reflect on Your Learning

Look back at the Learning Goals.
Which goal do you understand well enough
to be able to help a friend?
What would you do to help a friend?

Geometry

6

Building Castles

Learning Goals

- describe, name, and sort prisms
- construct prisms from their nets
- construct models of prisms
- identify, create, and sort symmetrical and non-symmetrical shapes
- draw lines of symmetry

Key Words

triangular prism

rectangular prism

net

symmetrical

line of symmetry

Look at the picture.
- Name the shapes and objects you see.
- How are some of the objects the same? Different?
- Where else might you find these shapes and objects?

Objects in Our World

Think about some of the objects you see around you.

What makes one object different from another?

Explore

➤ Go on a scavenger hunt.
  Find as many classroom items as you can that match each object below.
  Explain why each item matches an object.

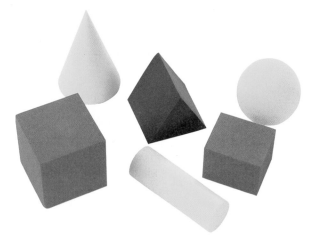

➤ Sort the items and objects. Record your sorting.

Show *and* Share

Share your sorting with another pair of students.
Discuss the ways you sorted the objects and items.
Find other ways to sort them.

Rectangular prism

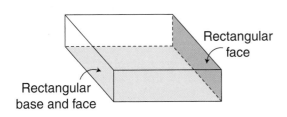

Rectangular face

Rectangular base and face

Triangular prism

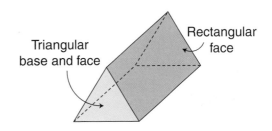

Triangular base and face

Rectangular face

Here are some ways to sort objects.

➤ You can sort by the shapes of the bases. Use these attributes.

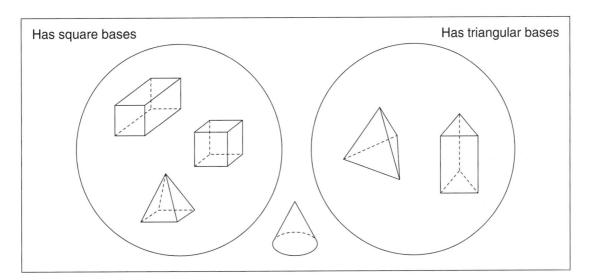

Has square bases

Has triangular bases

➤ You can sort by the shapes of the faces. Use these attributes.

Recall that "congruent" shapes match exactly.

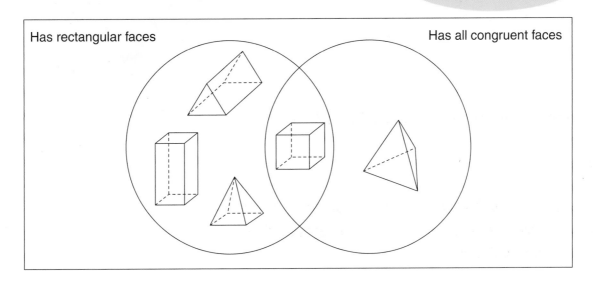

Has rectangular faces

Has all congruent faces

1. Name the prism that best represents each item.

a)

b)

c)

d)

2. Which attributes do all rectangular prisms have? Use these pictures to help you find out.

3. Which attributes do all triangular prisms have? Use these pictures to help you find out.

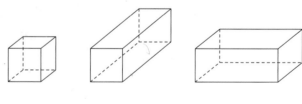

4. a) Sort these objects. Use these attributes: "Has triangular faces" and "Has two congruent faces" Record your sorting.

   b) Sort the objects again. Choose 2 different attributes to sort them by. Record your sorting.

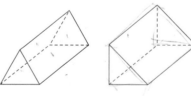

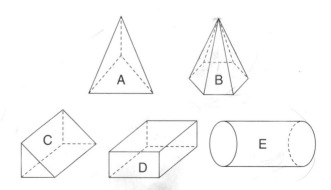

5. Look at the objects below.
Find 2 attributes.
Sort the objects.
Use the letters to help record your sorting.
Write your sorting rule.

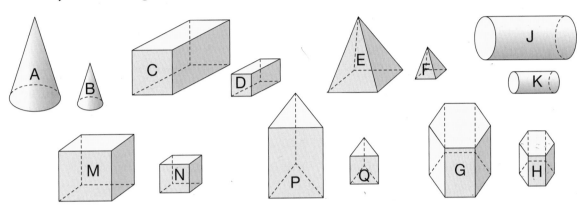

6. Use one of the words "all," "some," or "no" in place of □.
Copy and complete each sentence.
How do you know each sentence is true?
a) □ prisms have congruent bases.
b) □ rectangular prisms are cubes.
c) □ cubes are rectangular prisms.
d) □ prisms have all rectangular faces.
Show your work.

Patterning

There is a pattern in the numbers of edges, faces, and vertices of an object.
When you add the numbers of vertices and faces, the sum is 2 more than the number of edges.

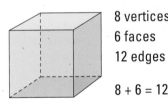

8 vertices
6 faces
12 edges

8 + 6 = 12 + 2

Reflect

How are a rectangular prism and a triangular prism alike?
How are they different?

Constructing Prisms

Explore ·

You will need Pattern Blocks.

➤ Use the orange square as the base of a prism.
Stack orange squares. Can you make a cube? Explain.
Continue to stack orange squares to make rectangular prisms.
Describe the prisms. How are they alike? How are they different?

➤ Use the green triangle as the base of a prism.
Make some different triangular prisms.
Describe the prisms. How are they alike? How are they different?

➤ Compare your rectangular prisms and your triangular prisms.
Describe how they are alike and how they are different.

Show *and* Share

Work with another group of students. Combine your Pattern Blocks.
Make a rectangular prism with a different size base.
How many different prisms can you make?
Make a triangular prism with a different size base.
How many different prisms can you make?

Sarah used modelling clay.
She began with a square base.
Sarah made each other face of her prism a square.
She made a cube.

Sarah added more clay to make her prism taller.
She made a square prism.

Sarah then started with a rectangular base.
She made a rectangular prism.

Sarah started with a triangular base.
She made a triangular prism.

1. Use modelling clay.
   Make 3 different objects.
   Each object should have at least one rectangular face.

2. Use modelling clay. Make an object that has each set of faces.
   a)

   b)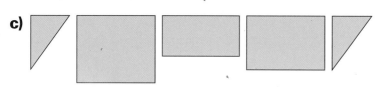

   c)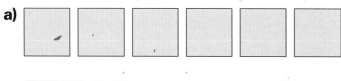

3. Two faces of a prism are shown.
   What could the prism be?
   How do you know?
   Which different prisms can you name each time?
   a)   b)    c)

4. Use modelling clay.

   Identify and make 2 different prisms for each description.
   a) It has 6 congruent square faces.
   b) It has 2 congruent triangle faces and 3 congruent rectangle faces.
   c) It has 3 pairs of congruent rectangle faces.

With which tool: modelling clay or Pattern Blocks can you make more prisms that are all different? Explain your choice.

Exploring Nets

Look at the two patterns below.

➤ Predict which pattern you think would fold to make a triangular prism.
   Give reasons for your prediction.

➤ Which prism do you think the other pattern would make when it is folded?
   How do you know?

Use the patterns provided by your teacher.
Cut out each pattern.
Fold it to make a prism.

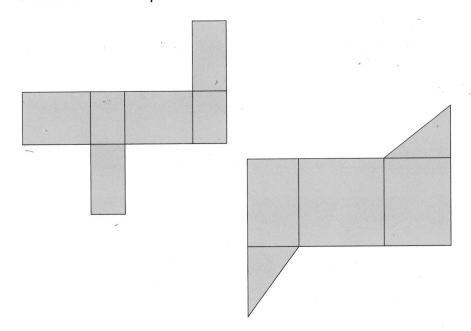

Show *and* Share

Compare your prisms with those of other students.
Were your predictions correct? Explain.
Explain to each other how you folded your pattern.

A pattern that can be folded to form an object is called a **net**.
A cereal box is a rectangular prism.
It can be made from a net.

A triangular prism can also be made from a net.

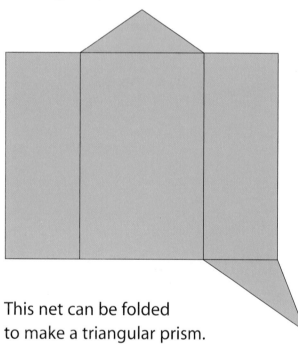

This net can be folded
to make a triangular prism.

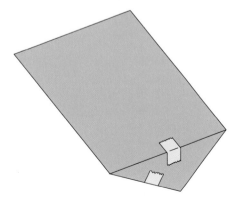

Your teacher will give you copies of the nets.

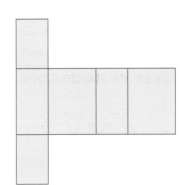

1. Use a large copy of this net.
   Colour the congruent faces the same colour.
   Fold the net to make an object.
   Where are the congruent faces on your object?

2. Which of these pictures are nets of a cube?
   How do you know?

   a) 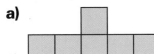   **b)** 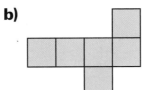   **c)**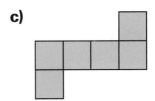

3. **a)** Which of the pictures below are nets of a triangular prism?
      How do you know?
   b) Predict which sides will meet when you fold the net.
      Cut out and fold each net to check.
   c) What do you notice about the lengths of the sides that join?

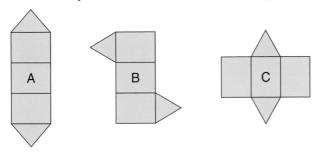

4. The net for a prism has 3 pairs of congruent rectangles.
   What kind of prism is it? How do you know?

Reflect

Suppose you see 6 rectangles in a picture.
How can you tell if it is a net of a rectangular prism?
Use words and pictures to explain.

Strategies Toolkit

Explore

Liam and Sophie have 36 Snap Cubes.
They use all the cubes to build a
rectangular prism.
How many different rectangular prisms
can they build?

Show *and* Share

Explain how you solved this problem.

Connect

Here is a similar problem.
How many different rectangular prisms can you
build with 24 Snap Cubes?

Understand

What do you know?
- There are 24 cubes.
- The cubes are used to build a
  rectangular prism.

Think of a strategy to help you solve
the problem.
- You can **work backward**.
- You know how many cubes
  you need. Use the cubes to make
  different rectangular prisms.

Plan

Strategies

- **Make a table.**
- **Use a model.**
- **Draw a picture.**
- **Solve a simpler problem.**
- **Work backward.**
- **Guess and test.**
- **Make an organized list.**
- **Use a pattern.**

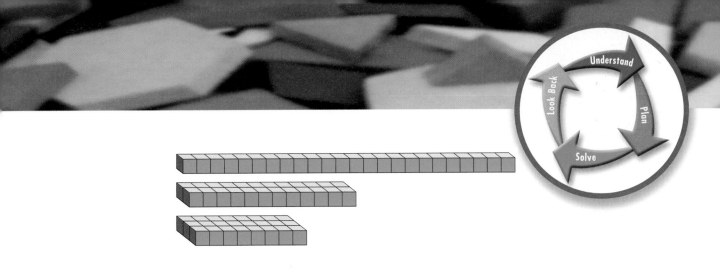

Solve

Start with a prism that is 1 cube high.
How many different prisms can you build?
Then try to build different prisms that are 2 cubes high,
3 cubes high, and so on.
How many different rectangular prisms did you make?
Record each prism that you made.

Look Back

How do you know you have found all possible
rectangular prisms?
Think of another way you could have solved this problem.

Practice

Choose one of the

Strategies

1. You have 100 Snap Cubes. How many larger cubes
   can you make using any number of the Snap Cubes?
   Record your work. What patterns do you see?

2. A rectangular prism is made with 9 Snap Cubes.
   How many different prisms can be made with these cubes?
   Suppose the number of cubes were doubled.
   How many different prisms can be made now?

Reflect

How can working backward help you solve a problem?
Use examples to explain.

Symmetrical Shapes

Each picture is **symmetrical**.

What do you think
"symmetrical" means?
How can you find out if
a shape is symmetrical?

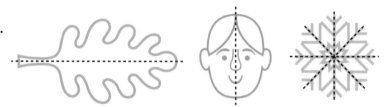

Explore

You need a copy of these shapes.
Use a Venn diagram with headings "Symmetrical" and "Non-symmetrical."
Sort the shapes.
Share the work.

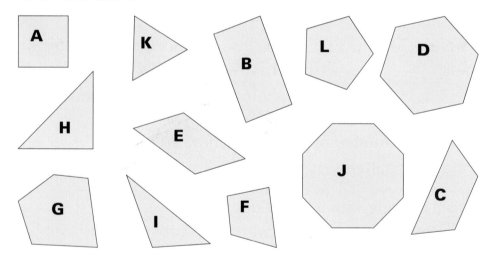

Show and Share

How did you identify symmetrical shapes?
Non-symmetrical shapes?
What do all symmetrical shapes have?

A **line of symmetry** divides a symmetrical shape into 2 congruent parts.

You can fold along the line and the 2 parts match.

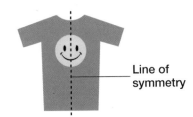

Line of symmetry

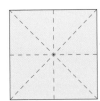

You can use a Mira to check a line of symmetry.

Some shapes have more than 1 line of symmetry.

A square has 4 lines of symmetry.

Some shapes are non-symmetrical. They have no line of symmetry.

Practice

1. Which pictures are symmetrical? Non-symmetrical? How do you know?

a)    b)   c)    d)

2. Write your name in capital letters.
   a) How many letters have 1 line of symmetry?
   b) Are these letters more than one-half of your name?
      Less than one-half?

3. Which flags below have each symmetry?
   How do you know?
   • a vertical line of symmetry
   • a horizontal line of symmetry
   • more than 1 line of symmetry
   • no lines of symmetry
   a) Canada   b) The United States   c) Japan
                                of America

   d) France   e) China   f) New Zealand

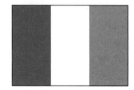  

4. On grid paper, draw a symmetrical picture.
   Tell how you know it is symmetrical.
   Draw a non-symmetrical picture.
   Tell how you know it is not symmetrical.

 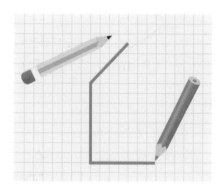

Reflect

When you see a picture, how can you tell if it is symmetrical?

Line Symmetry

The Lakota Morning Star quilt represents the planet Venus in the sky. Lakota women began making quilts in the late 19th century. Buffalo were scarce and quilts replaced buffalo robes. A star quilt is given as a gift on important occasions.

Explore

You will need grid paper, Pattern Blocks, and a Mira.

➤ Fold the grid paper in half.
  Use Pattern Blocks.
  Make a design on one side
  of the fold line.
  Your design must touch
  the fold line.
  Trace around your design.
  Do not draw on the fold line.

➤ Open the paper.
  Use the Mira.
  Make a mirror image of
  the design on the other side
  of the fold line.
  Trace around the mirror image.
  Remove the Pattern Blocks.

➤ Find any lines of symmetry
  on the picture.

LESSON FOCUS | Construct symmetrical shapes and draw lines of symmetry.

Show *and* Share

Show your work to another pair of students.
How did you find the lines of symmetry?
What did you notice about the fold lines?

Connect ·

Here is one way to make a symmetrical shape.

➤ Fold a piece of paper.
Draw a shape.
Use the fold line as one side of the shape.
Cut out the shape.

Recall that a symmetrical shape has one or more lines of symmetry.

➤ Unfold the paper.
The fold line is a line of symmetry.

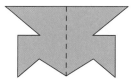

Practice ·

1. How many lines of symmetry does each Pattern Block have?
   Trace each block and draw the lines of symmetry.

   a)

   b)

   c)

   d)

   e)

   f)

2. Trace the shapes that have line symmetry.
Draw the lines of symmetry.

a)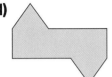

b)

c)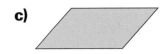

d)

e)

f)

3. Choose a shape in question 2 that has line symmetry.
Fold a piece of paper in half.
Use scissors.
Cut out a shape so that
when the paper is unfolded, it matches
the shape you chose in question 2.

4. One-half of a symmetrical shape is shown.
The broken line is a line of symmetry.
Copy the shape and the line of symmetry on dot paper.
Complete the shape.

a)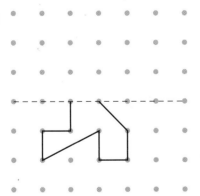

b)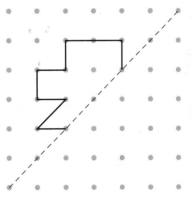

5. Work with a partner.
Use dot paper.
Take turns to draw a symmetrical shape.
Explain how you know your partner's shape is symmetrical.
Draw the lines of symmetry on your partner's shape.

6. Work with a partner.
Use a geoboard and geobands.
Divide the geoboard in half.
The dividing line is a line of symmetry.
Make one-half of a shape on the geoboard.
Trade geoboards.
Make the other half of your partner's shape.
Trade geoboards again to check that
the shapes are symmetrical.

7. This shape does not have a line of symmetry.
Copy the shape on grid paper.
Add one or more squares so it has
a line of symmetry.
How many different ways can you do this?
Record each way on grid paper.

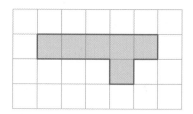

8. Use Pattern Blocks to make a star quilt design.
Try making three different star designs.
How many lines of symmetry does each design have?

9. Fold a piece of paper in half.
Cut the paper to make each shape below,
when the paper is unfolded.
  a) a triangle
  b) a quadrilateral
  c) a pentagon
  d) a hexagon
Tell about the shapes you created.

Reflect

How well do you think you understand symmetry?
Explain how to construct a shape that has line symmetry.

Sorting by Lines of Symmetry

A line of symmetry divides a shape into two congruent parts. When the shape is folded along its line of symmetry, the parts match exactly.

How can you tell that the two parts are congruent?

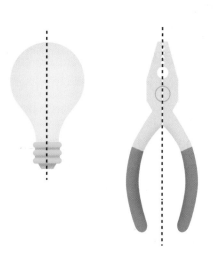

You will need a copy of the shapes below and a Mira.

➤ Draw as many lines of symmetry on each shape as you can find.

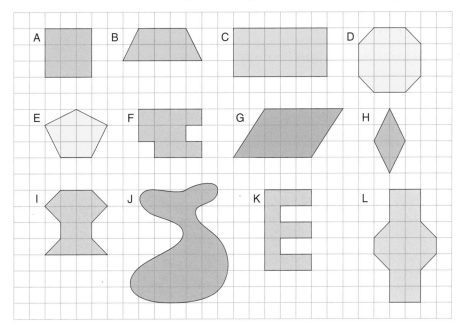

➤ Draw a Venn diagram.
Label the circles: "0 lines of symmetry;" "1 line of symmetry;" and "More than 1 line of symmetry."
Sort the shapes.

Show *and* Share

Share your sorting with a classmate.
Explain how you know which shapes are symmetrical.

Connect

A shape has symmetry when a line of symmetry can be drawn on it.
Some shapes have no lines of symmetry.
Some shapes have more than one line of symmetry.

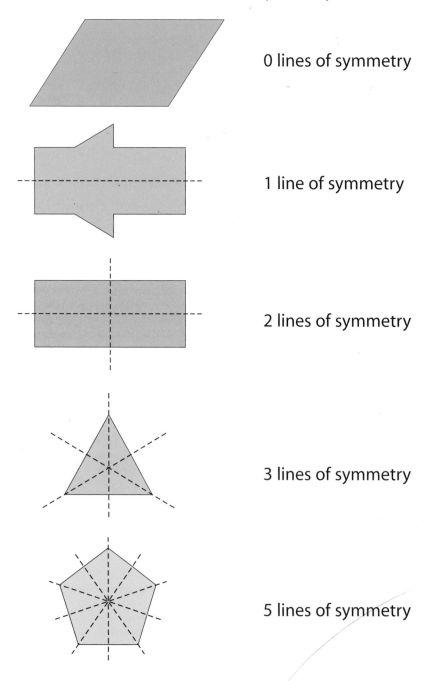

0 lines of symmetry

1 line of symmetry

2 lines of symmetry

3 lines of symmetry

5 lines of symmetry

Use a Mira when it helps.

1. Is each broken line a line of symmetry? How do you know?

a) 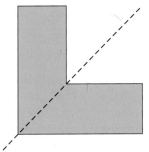 b) 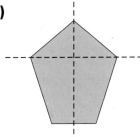 c)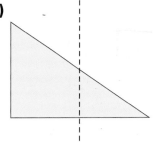

2. a) How many lines of symmetry does each triangle have?

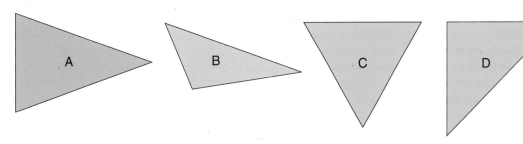

b) Which triangle has the most lines of symmetry?
   Describe the triangle.

3. a) How many lines of symmetry
   does each regular polygon have?

All sides of a regular polygon are equal.

b) How many sides does each regular polygon have?
c) Predict the number of lines of symmetry
   for a regular polygon with 10 sides, a *decagon*.
   Explain your thinking.

4. Use a copy of this diagram. Sketch a shape in each cell.

Shape has …	0 lines of symmetry	1 line of symmetry	More than 1 line of symmetry
3 sides			
4 sides			
5 sides			
6 sides			

5. a) How would you change this design to make it symmetrical about the line of symmetry shown?

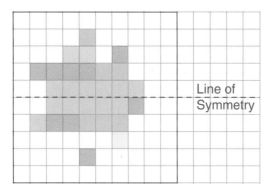

Line of Symmetry

b) How many different ways could you change this design but still make it symmetrical?
Use grid paper to show the different ways.

At Home

Reflect

When you see a shape, how can you identify how many lines of symmetry it has?
How do you know you have found all the lines of symmetry?

Look through magazines to find pictures with symmetry. Cut out the pictures. Use a ruler to draw each line of symmetry.

What's My Rule?

You will need a set of *What's My Rule?* game cards, scissors, 2 labels, and two 1-m lengths of string.

Cut out the game cards.
Spread them out, face up.
Use the string to make 2 loops.
Label one loop "Matches"
and one loop "Discards."

➤ Player A thinks of a secret rule that describes some of the shapes on the cards. The rule could be:
  • it has more than 1 line of symmetry; or
  • it has all sides equal; or
  • it is a square

➤ Player A chooses 2 game cards. One card must fit the rule.
  He places it face up inside the "Matches" loop.
  The other card must not fit the rule.
  He places it face up in the "Discards" loop.

➤ Player B chooses a game card.
  If she thinks the card fits the rule, she places it inside the "Matches" loop. Otherwise, she places it in the "Discards" loop.

➤ Player A tells Player B whether she is correct.
  If she is correct, she can guess the rule.
  If she is not correct, she cannot make a guess.

➤ Players C and D continue until someone guesses the secret rule.

➤ Switch roles. Another player thinks of a secret rule.
  The other players take turns trying to guess the new rule.
  The winner takes the fewest turns to guess the rule.

LESSON

1. Look at the objects below.
   Choose 2 attributes.
   Draw a Venn diagram.
   Sort the objects.
   Record your sorting.

   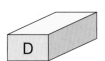

2. Use the pictures above.
   a) Identify a triangular prism.
      Write all that you know about it.
   b) Identify a rectangular prism.
      Write all that you know about it.

3. Use modelling clay.
   Make two different triangular prisms.
   How are the prisms alike? How are they different?

4. Use 12 orange Pattern Blocks.
   How many different rectangular prisms can you make?
   How are the prisms alike? How are they different?

5. Which pictures below show a net for a rectangular prism?
   How could you check?

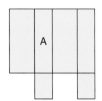 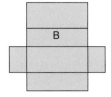 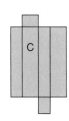

6. Is each shape symmetrical?
How do you know?

a)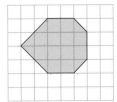
b)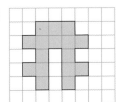
c)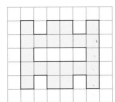
d)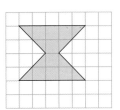

7. Copy each shape in question 6 on grid paper.
Draw all the lines of symmetry.

8. One-half of a symmetrical shape is shown.
The broken line is a line of symmetry.
Copy the shape and line of symmetry
on dot paper.
Complete the shape.

9. Sketch each shape. Draw its lines of symmetry.

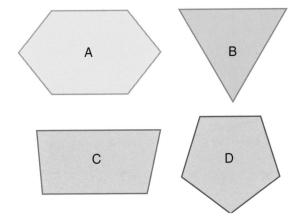

10. Draw a Venn diagram.
Sort the shapes in question 9.
Use these attributes:
"Has 0 lines of symmetry;"
"Has 1 line of symmetry;"
"Has more than 1 line of symmetry"

UNIT

6 **Learning Goals**

☑ describe, name, and
sort prisms
☑ construct prisms from
their nets
☑ construct models of prisms
☑ identify, create, and sort
symmetrical and
non-symmetrical shapes
☑ draw lines of symmetry

Building Castles

Stefanie wants to make
a castle with cubes.
First she has to construct
the cubes.

Part 1

Stephanie has this set of pentominoes.

A *pentomino* is an
arrangement of
5 congruent squares.

Which pentominoes are symmetrical?
Your teacher will give you a copy of the pentominoes.
Draw the lines of symmetry.

Stefanie knows that some of these
pentominoes will fold to make an open cube.
Which pentominoes above can be folded
to make an open cube?

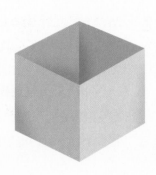

Part 2

Stefanie will use the pentominoes that fold to make open cubes. She will draw a square on each pentomino so when it folds, it makes a closed cube. Draw a square on each pentomino so it is a net of a cube.

Each net is a hexomino because it has 6 squares. Draw the hexominoes on grid paper. Which hexominoes are symmetrical? Draw the lines of symmetry.

Part 3

Combine any pentomino and hexonimo to make a new shape that is symmetrical. Sketch this new shape. Draw the line of symmetry. Is it possible to combine two shapes to make a new shape with more than one line of symmetry? Investigate to find out.

A *hexomino* is an arrangement of 6 congruent squares.

Reflect on Your Learning

Which learning goals have you met?
Find examples of your work that show how you met these goals.

UNIT

1

1. a) Use Colour Tiles to build this pattern.

Figure	Tiles in a Figure
1	2
2	5
3	8
4	11
5	14
6	17

b) Draw the pattern on grid paper.

c) Write a pattern rule.

d) How many tiles would be in the 8th figure?
How did you find out?

2

2. Add or subtract. How do you know each answer is reasonable?

a) $\begin{array}{r} 3057 \\ -\ 1999 \end{array}$ b) $\begin{array}{r} 3210 \\ +\ 5909 \end{array}$ c) $\begin{array}{r} 6666 \\ -\ 3777 \end{array}$ d) $\begin{array}{r} 8376 \\ +\ 1234 \end{array}$

3

3. a) What is the product when you multiply a number by 0?
Draw a diagram to explain your answer.

b) What is the product when you multiply a number by 1?
Explain how you know.

4

4. Show each time on a digital clock and an analog clock.

a) five forty-five

b) ten thirty

c) 3 minutes after 7

5. Show each time on a 24-hour digital clock.

a) six thirty-two in the afternoon

b) eleven thirteen at night

6. Use 1-cm grid paper. Draw a large shape.
Find the area of your shape.
Explain how you did this.

7. Use 1-cm grid paper.
Draw as many rectangles as you can that have area 24 cm².

8. Fold a paper strip to show fifths.
Colour $\frac{3}{5}$ of the strip.
What fraction of the strip is not coloured?

9. Use 18 counters.
   a) What fraction of the set is 9 counters?
   b) How many counters are in $\frac{5}{6}$ of the set?
   c) What other fractions can you show with 18 counters?

10. Order each set of fractions from least to greatest.
   a) $\frac{1}{5}, \frac{1}{3}, \frac{1}{7}$ **b)** $\frac{6}{8}, \frac{3}{8}, \frac{1}{8}$ **c)** $\frac{4}{9}, \frac{4}{7}, \frac{4}{12}$

11. Use a hundredths grid to represent 1.
Colour a grid to show each fraction.
   a) $\frac{11}{100}$ **b)** $\frac{4}{10}$ **c)** $\frac{8}{10}$ **d)** $\frac{73}{100}$

12. Write an equivalent decimal for each number.
Use a hundredths grid to show how you know
the decimals are equivalent.
   a) 0.40   **b)** 0.2   **c)** 0.9   **d)** 0.70

6

13. Look around your classroom.
Find 4 rectangular prisms.
What is the same about all the prisms?

14. Use grid paper.
Draw a symmetrical shape.
   a) How do you know the shape is symmetrical?
   b) How many lines of symmetry does it have?

Data Analysis

7

Using Data to Answer Questions

Books Read in April

Angie	📖 📖 📖 📖 📖 📖 📖
Shane	📖 📖 📖
Paul	📖 📖 📖 📖
Jason	📖 📖 📖 📖 📖
Deanna	📖 📖 📖 📖 📖 📖

📖 Represents 1 book

Lunch	Number of Students
Pizza	⊞⊞ IIII
Burger and fries	⊞⊞ III
Tuna sandwich	IIII
Taco	⊞⊞
Submarine	⊞⊞ I

Learning Goals

- interpret graphs
- compare pictographs with different keys
- compare bar graphs with different scales
- draw pictographs and bar graphs
- compare pictographs and bar graphs

Key Words

key

scale

vertical bar graph

vertical axis

horizontal axis

horizontal bar graph

trial

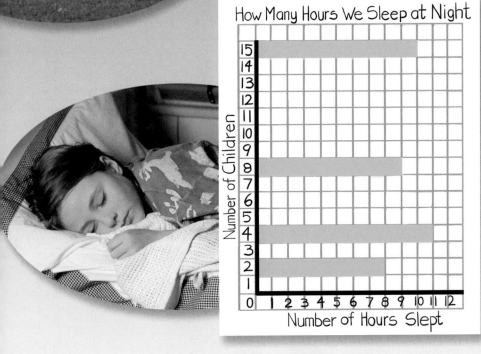

Favourite Sports of 100 Children	
Hockey	32
Soccer	27
Lacrosse	9
Baseball	22
Basketball	10

How Many Hours We Sleep at Night

These graphs and charts show data about Grade 4 students.

What can you find out from each graph and chart?

LESSON

1

Reading Pictographs and Bar Graphs

Explore

These pictographs show data from a Grade 4 reading group.

The pictograph on the left is from the Unit Launch.
The pictograph on the right shows the same data.

How are the pictographs the same? Different?
Who read the most books in April?
How do you know?

Show and Share

Compare your answers with those of other classmates.
Which pictograph did you use to answer the questions?
Take turns to ask each other questions about
the pictographs.

The title of a graph tells you what the graph shows.
The labels tell you what data are displayed.
In a pictograph, symbols are used to show data.

The **key** shows what each symbol represents.
For the pictograph on the right in *Explore*,
the key is 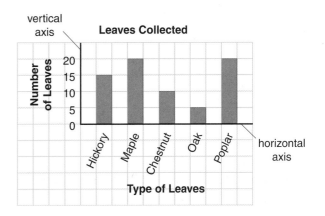 represents 2 books.

In a bar graph, bars are used to show data.
The numbers show the **scale**.
These bar graphs show the same data.

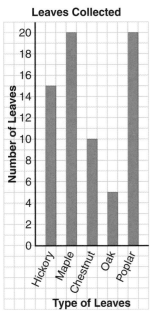

In this graph,
1 square represents 1 leaf.

The bar graphs have different scales.
The bars are shorter in the graph on
the right.
A scale is chosen so the size of the
graph is manageable.

In this graph,
1 square represents 5 leaves.

These are **vertical bar graphs**.
The bars are drawn upward.
The numbers are on the **vertical axis**.
The types of leaves are on the
horizontal axis.

1. This pictograph shows the types of gym shoes Grade 4 students wear at Zeina's school.

   a) What is the key?

   b) What is the most common type of gym shoe?

   c) How many students wear basketball shoes?

   d) Which 2 types of shoes are worn by the same number of students? How do you know?

Types of Gym Shoes

Basketball	
Trail/Hiking	
Cross Trainers	
Court	
Running	

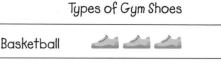

represents 3 students

2. This bar graph shows the typical number of sunny days each year for 6 cities.

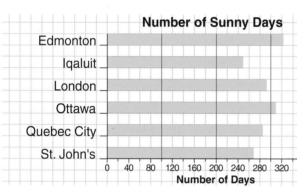

   a) What is the scale?

   b) Which cities have more than 300 sunny days?

   c) Which cities have between 200 and 300 sunny days?

   d) Which city has the most sunny days?

   e) Suppose the numbers on the axis were not given. Could you still answer part d? Explain.

   f) How many sunny days does Iqaluit have? Is your answer exact or approximate? Explain.

3. The graph in question 2 is a **horizontal bar graph**. How is this graph like the graphs in *Connect?* How is it different?

4. These graphs show the number of goals scored by 6 NHL players in the 2005/2006 season.

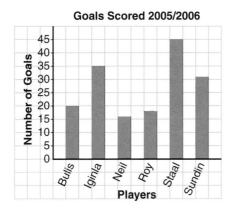

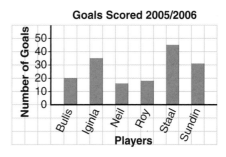

a) Who scored the most goals? The fewest goals?
b) Which player scored about one-half the goals of Iginla?
c) Which graph was easier to use, to answer parts a and b? Explain.
d) What is the scale on each graph?
e) Which scale do you think is more appropriate? Explain.

5. a) What does this graph show?

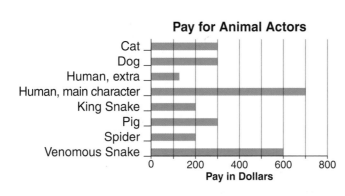

b) Which animal actors receive the same pay?
c) Which actors' pay is one-half that of a venomous snake's?
d) Why do humans appear twice in the graph?
   What is the difference in their pay?
e) Why do you think some animal actors get paid more than others?

Reflect

How does the key affect the appearance of a pictograph?
How does the scale affect the appearance of a bar graph?

Drawing Pictographs

You will need a bag of 20 two-colour counters.

➤ Empty the counters onto a desk.

➤ Choose one colour. Make a table.
Record the number of counters showing this colour.
This is your first **trial**.

➤ Return the counters to the bag. Do 4 more trials.
Count the same colour each time.
Record the results of each trial.

Trial	Number of red counters
1	
2	
3	
4	
5	

➤ Graph your data in a pictograph. Use a key.
Each symbol should represent more than 1 counter.

➤ What do you know from looking at the pictograph?

Show *and* Share

Did you have fractions of symbols in your pictograph? Explain.
Could you have chosen a key so there would be no fractions? Explain.

Aliyah asked Grades 4 and 5 students in her school
how they travel to school each day.
Here are her results:

Type of transportation to School	Walk	Bus	Bike	Car
Number of Students	30	45	5	25

Aliyah chose 人 for the symbol because she collected data
on the number of students.

To make sure her graph was not too large,

Aliyah chose 人 to represent 10 students.

Then, 𝇋 represents 5 students.

To show 30 students, Aliyah needed 3 symbols: 人人人

To show 45 students, Aliyah needed 4 symbols and $\frac{1}{2}$ a symbol: 人人人人𝇋

To show 5 students, Aliyah needed $\frac{1}{2}$ a symbol: 𝇋

To show 25 students, Aliyah needed 2 symbols and $\frac{1}{2}$ a symbol: 人人𝇋

To draw the pictograph, Aliyah wrote each type of transportation on the vertical axis.
Then, she drew the correct number of symbols beside each type of transportation.
Aliyah completed the pictograph with a key, a label on the axis, and a title.

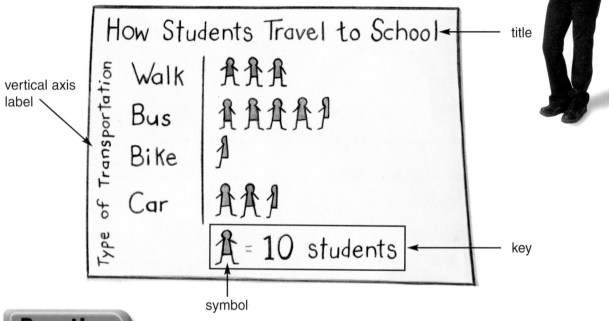

title

vertical axis
label

How Students Travel to School

Type of Transportation

Walk
Bus
Bike
Car

🧍 = 10 students

key

symbol

Practice

1. Each table has data for a pictograph.
   Suppose you drew each pictograph. What key would you use? Why?

a)
Favourite Type of Movie	Number of Students
Action	6
Comedy	12
Drama	8
Horror	2
Mystery	4

b)
Favourite Colour	Number of People
Red	15
Yellow	25
Blue	10
Green	25
Orange	40

c)
Favourite Family Activity	Number of Students
Hunting	60
Fishing	50
Dog-sledding	80
Trapping	70
Camping	100

2. a) Draw a pictograph to display these data.

Students in the Band

Grade	4	5	6	7	8
Number of Students	9	5	6	11	13

b) How did you choose your key?

c) Write what you know about the band.

3. a) Draw a pictograph to display these data.

Time When People Take the Bus in the Morning

Time of Day	6:00	7:00	8:00	9:00	10:00	11:00	12:00
Number of People	4	8	14	2	5	8	10

b) What do you know from the pictograph?

c) Write two questions using the data from the graph.
Exchange questions with a classmate.
Answer your classmate's questions.

4. This table shows the typical number of eggs some animals lay.

a) Draw a pictograph.
How did you choose your key?

b) A seahorse lays about 200 eggs.
How would you include this data
on your pictograph?
Would you need to change anything?
Explain the change.
Show your work.

ANIMAL	NUMBER OF EGGS
FROG	60
PYTHON	25
SALAMANDER	60
TURTLE	95

Reflect

When you draw a pictograph, how do you decide what
key to use?
Use words, pictures, or numbers to explain.

Drawing Bar Graphs

This table shows the typical
life spans of some animals.

Use grid paper. Draw a bar graph.
Choose a suitable scale.

Animal	Typical Life Span
Asian elephant	40 years
Black rhinoceros	15 years
Killer whale	65 years
Polar bear	20 years

Show *and* Share

Share your graph with another
pair of students.
How are your graphs the same? Different?

The typical life span of a Galapagos tortoise is 100 years.
Suppose you wanted to add the life span of this tortoise to your graph.
Discuss how you might have to change your graph.

Connect

Bar graphs may be drawn vertically or horizontally.
You can graph the data below on grid paper.

Animal	Typical Life Span
Bottle-nosed dolphin	40 years
Brown bear	22 years
Fin whale	85 years
Potbelly seahorse	8 years

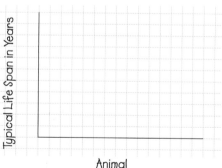

Draw 2 axes. Label the horizontal axis "Animal."
Label the vertical axis "Typical Life Span in Years."

There are 20 squares along one side of the 1-cm grid paper.

If we count by 1s, the bar will go only to 20. The greatest number in the table is 85.

If we count by 2s, the bar will go only to 40.

➤ Count by 5s for the scale. The scale is 1 square represents 5 years.

➤ Draw a vertical bar for each animal in the table. Estimate the lengths of the bars for 22 years and 8 years. The bar for 22 years ends less than halfway between 20 and 25. The bar for 8 years ends slightly more than halfway between 5 and 10.

➤ Write a title for the graph.

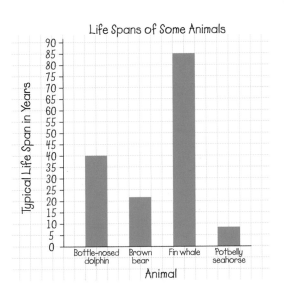

Life Spans of Some Animals

Practice

1. The number of wins in 2005 is shown for 4 Major League Baseball teams.
   a) Draw a bar graph to show the number of wins.
      For the scale, count by 5s.
   b) Write two things you know from looking at your graph.

Teams	Number of Wins
Blue Jays	80
Mariners	69
Tigers	71
Yankees	95

2. Stefan had a bag of coloured candy.
He counted each colour.

Stefan's Candy

Colour	Tally	Number
Brown	ЖЖ	5
Red	ЖЖ ЖЖ II	12
Yellow	ЖЖ ЖЖ	10
Blue	ЖЖ I	6
Orange	ЖЖ IIII	9
Green	ЖЖ II	7

a) Draw a bar graph.
Use a scale of 1 square represents 1 candy.
b) Draw a bar graph.
Use a scale of 1 square represents 2 candies.
c) Which scale is better for this graph? Explain.

3. The children in l'école Orléans
estimated the time they took
to get to school.
a) Draw a bar graph.
Explain why you chose the scale
you did.
b) Compare your graph with a
classmate's graph.
Do both graphs match? Explain.
c) How many children take the
greatest time?
d) Some children take only 5 minutes.
Do they live closest to the school?
Explain your answer.

**How Long It Takes
to Get to School**

Time in Minutes	Number of Children
5	11
10	22
15	38
20	37
25	45
30	10
35	33
40	20
45	12

4. This table shows the heights of some players from the 2006 Canadian women's Olympic hockey team.

a) Draw a bar graph to display these data.

b) Explain why each of these parts of your graph is important: title, bars, labels, scale

c) Why is your scale *not* 1 square represents 1 cm?

Name	Height
Apps	183 cm
Botterill	175 cm
Kellar	170 cm
Ouellette	180 cm
Piper	165 cm
St. Pierre	175 cm
Wickenheiser	178 cm

5. a) Draw a bar graph. Explain your choice of scale.

b) Which city had the fewest wet days?

c) Why do you think Victoria had more wet days than Edmonton?

d) Write a question you can answer using the table or the graph. Answer the question.

Show your work.

City	Typical Number of Wet Days Each Year (1961–1990)
Charlottetown	177
Edmonton	123
Fredericton	156
Montreal	162
Ottawa	159
Victoria	153

Math Link

Your World

The driest town in Canada is Osoyoos, B.C. It typically receives less than 20 cm of rainfall each year.

In the Inkaneep native dialect, the name Osoyoos means "where the water narrows."

Reflect

How did you use what you know about reading bar graphs to draw a bar graph?

Comparing Pictographs and Bar Graphs

These 2 graphs show the same data.

Videos Rented in One Store on One Day

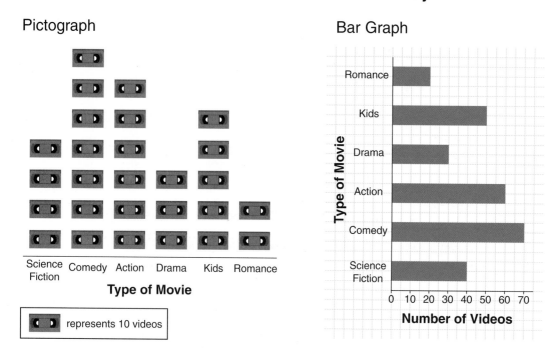

Pictograph

Science Fiction · Comedy · Action · Drama · Kids · Romance
Type of Movie

Bar Graph

represents 10 videos

Look at each graph.
List all the things you know from looking at each graph.
How are the graphs the same? Different?

Show *and* Share

Share your list with another pair of classmates.
Which type of movie is most popular? Least popular?
Which graph is easier to read? Why?

Fun Times Park rents equipment.
This pictograph shows the equipment
rentals for one week in July.

This bar graph shows the
Saturday activities for one
Saturday in July.

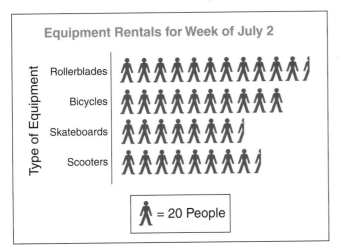

Pictographs and bar graphs are similar.
In a pictograph, symbols show the data.
In a bar graph, bars show the data.

From both the pictograph and the bar graph,
we can only estimate the number of people.
It is usually easier to estimate the number from a bar graph.
We use the scale to do this.

A pictograph has more impact; it is visually appealing.
In a pictograph, we use the key to help estimate numbers.

Practice

1. This table shows the after-school
   activities of some students.
   a) Which activity was chosen by
   the most students? The fewest?
   b) Would you use a bar graph or
   pictograph to display these data?
   Explain.
   c) Draw the graph you chose in part b.
   d) Do you think the data would be the same in your school? Explain.

Activity	Number of Students
Music lessons	18
Dancing lessons	24
Playing sports	36
Swimming lessons	60
Computer club	42

2. Some children were asked to name their favourite animal.

    a) How many children like dogs?

    b) List the animals from most popular to least popular.

    c) How many children were asked? How do you know?

    d) Suppose you had to draw a bar graph to show these data. How could you use the key to help you decide the scale?

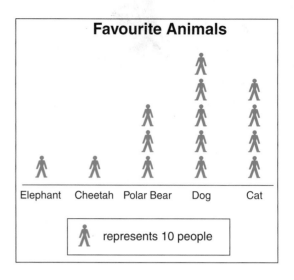

Favourite Animals

Elephant   Cheetah   Polar Bear   Dog   Cat

represents 10 people

3. Children from three Grade 4 classes were asked to tell their eye colour.

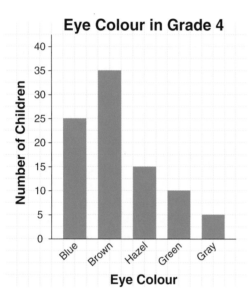

Eye Colour in Grade 4

    a) Which eye colour is most common? Least common?

    b) How many more blue eyes are there than hazel?

    c) How many more brown eyes are there than green?

    d) What is the scale?

    e) Suppose you had to draw a pictograph to show these data. How could you use the scale to help you decide the key?

    f) Make up your own question about this graph. Trade questions with a classmate. Answer the question.

4. a) What does the bar graph show?
  b) Which vegetable takes the longest time to grow?
  c) What else do you know from the graph?
  d) Suppose you wanted to display these data as a pictograph. What key would you use? How many symbols would you need for each vegetable?

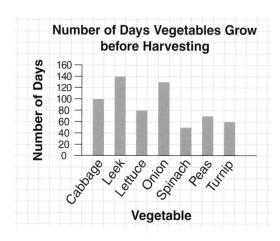

5. Scott found how many people in Grades 1 to 6 wear glasses. Here are his data.

Grade	Children Who Wear Glasses
1	15
2	5
3	25
4	40
5	30
6	10

a) Draw a pictograph.
b) How did you choose the key?
c) Draw a bar graph.
d) How did you choose the scale?
e) Suppose you wanted the bar graph to fill a page. What scale would you use? Explain.

At Home

Reflect

Which do you find easier to read: a pictograph or a bar graph? Explain.

Look through newspapers and magazines.
Find a bar graph.
What is its scale?

Strategies Toolkit

Explore ∙∙

The Grade 4 music class has 26 students.
Each student plays the clarinet, recorder, or trumpet.
There are 12 boys in the class.
Of the 8 students who play the recorder, 5 are girls.
Three boys play the trumpet.
Eight students play the clarinet.
How many girls and boys play each instrument?

Show *and* Share

Describe the strategy you used to solve the problem.

Connect ∙∙∙

Strategies

- **Make a table.**
- **Use a model.**
- **Draw a picture.**
- **Solve a simpler problem.**
- **Work backward.**
- **Guess and test.**
- **Make an organized list.**
- **Use a pattern.**

At a track and field meet, 42 students won medals.
- 18 medals were won in field events.
- 9 gold medals were won in field events.
- 15 silver medals were won in track events.
- 10 bronze medals and 14 gold medals were won.
How many gold, silver, and bronze medals were
won in track events? In field events?

What do you know?
- Some of the data are given above.
- Use those data to find the unknown data.

Think of a strategy to help you solve the problem.
- You can **make a table**.
- Fill in what you know. Use addition and subtraction
  to find the missing numbers in the table.

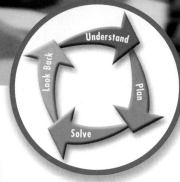

Copy and complete the table.

Medals	Gold	Silver	Bronze	Total
Track		15		
Field	9			18
Total	14		10	42

How many of each type of medal was won?

How do you know your answers are correct?
How could you have solved this problem another way?

Practice

Choose one of the
Strategies

1. There are rainy days, sunny days, and cloudy days.
   - September and October had the same number of rainy days.
   - There were 6 cloudy days in September.
   - There were 10 rainy days in total for both months.
   - There were 3 more cloudy days in October than in September.
   How many sunny days were there in September? In October?

2. Mr. Chu's class counted animals on its field trip.
   How many of each type of animal were seen in the woods?
   In the stream?
   - 30 animals were counted. There were 16 animals in the woods.
   - 2 omnivores were in the stream, and 4 omnivores were seen in total.
   - In the stream, there were 3 times as many herbivores as omnivores.
   - There were half as many carnivores in the stream as in the woods.

Reflect

How can a table help you solve a problem?
Use words and numbers to explain.

1

1. **a)** What does this bar graph show?
   b) What is the scale?
   c) Write 3 things you know from the graph.

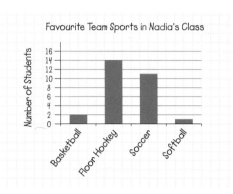

Favourite Team Sports in Nadia's Class

Number of Students

Basketball Floor Hockey Soccer Softball

2

2. Madhu found out how many children in her school watched the Canadian hockey team play. The team won a gold medal.
   a) Draw a pictograph. Which key did you use? Explain your choice.
   b) Choose a different key. Draw another pictograph.
   c) Compare the two pictographs. Which is easier to use to answer these questions?
   - How many more Grade 4 children watched the game than Grade 2 children?
   - How many children watched the game?

Grade	Children Who Watched the Game
1	25
2	40
3	35
4	55
5	65
6	50

3

3. The table shows the number of people who like to hold each animal at the zoo.

Animal	Number of People
Banana slug	9
Gila monster	22
Koala	32
Macaw	14
Monkey	16
Rosy boa	6

a) Draw a bar graph. What scale did you use?
b) Draw a different bar graph to display these data. How did you choose your scale this time?

c) How many more people prefer to hold a koala than a monkey?

d) Which animal is most popular? Least popular?

e) Which graph was easier to use to answer the questions in part d? Explain your choice.

f) Write your own question about the bar graphs. Answer the question.

4. Look at this pictograph.

Life Spans of Birds in Captivity	
Cockatoo	❤ ❤ ❤ ❤ ❤ ❤ ❤ ❛
Rhea	❤ ❤ ❤ ❤
Vulture	❤ ❤ ❤
Ostrich	❤ ❤ ❤ ❤
Swan	❤ ❤ ❤ ❤ ❤
Bald Eagle	❤ ❤ ❤ ❤ ❤

❤ represents 10 years

a) Find two birds whose combined life spans are less than that of a cockatoo.

b) A canary's life span is 25 years.
How would you show 25 years on this graph?

c) Which bar graph below shows the same data as the pictograph above? How do you know?

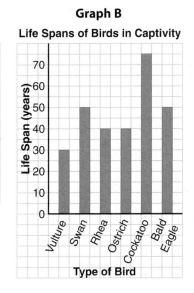

Graph A

Life Spans of Birds in Captivity

Graph B

Life Spans of Birds in Captivity

UNIT

7 Learning Goals

- ☑ interpret graphs
- ☑ compare pictographs with different keys
- ☑ compare bar graphs with different scales
- ☑ draw pictographs and bar graphs
- ☑ compare pictographs and bar graphs

d) Which graph do you think best displays the data? Give reasons for your choice.

Using Data to Answer Questions

Part 1

Two students collected these data from students in Grades 4 and 5.

Colours of Eyes in Grades 4 and 5

Blue	Brown	Green	Other
30	42	18	10

Each student drew a graph.

Emil's graph

Safia's graph

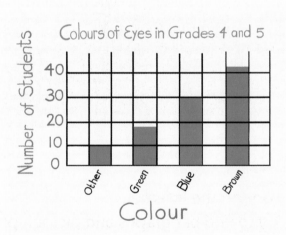

- How are the graphs the same? Different?
- What is the scale for each graph?
- Why do you think each person chose the scale he or she did?
- Why do you think the scale is *not* 1 square to 1 person?
- Choose a different scale or choose a key.
  Draw your own graph.
  Justify your choice of scale.
- Write 3 things you know from your graph.

Part 2

Look through newspapers and magazines.
Try to find bar graphs and pictographs.
Sketch each graph you find.
Identify its scale or key.

Part 3

Look on the Internet.
Try to find bar graphs and pictographs.
Print each graph you find.
Identify its scale or key.

Reflect on Your Learning

Tell what you now know about bar graphs and pictographs that you did not know at the beginning of the unit.

Find two examples where pictographs or bar graphs are used outside the classroom.

Multiplying and Numbers

At the Garden Centre

Learning Goals

- use personal strategies to multiply
- estimate products
- use models and arrays to multiply and divide
- multiply a 2-digit and a 3-digit number by a 1-digit number
- estimate quotients
- divide a 2-digit number by a 1-digit number
- use personal strategies to divide
- relate multiplication and division
- identify patterns in multiplication and division

Dividing Larger

Key Words

- multiplication sentence
- quotient
- remainder
- division sentence

May-Lin works in a garden centre. She has planted seeds that grow into seedlings.

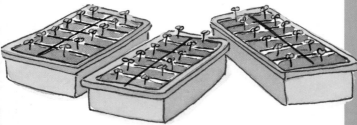

- How many seedlings are there?
- How did you find out?
- How many different ways can you find the answer?

May-Lin will replant the seedlings into other boxes like this:

- How many of these boxes does May-Lin need for all the seedlings?
- How do you know?

1 Exploring Multiplication Patterns

Explore

You will need a calculator.

➤ Use a calculator to find each product.

4 × 1	9 × 1	5 × 5	2 × 3
4 × 10	9 × 10	5 × 50	2 × 30
4 × 100	9 × 100	5 × 500	2 × 300

What patterns do you see?

➤ Use patterns to find each product.
Check with a calculator.

7 × 1	8 × 1	4 × 2	2 × 9
7 × 10	8 × 10	4 × 20	2 × 90
7 × 100	8 × 100	4 × 200	2 × 900

> 30 is a multiple of 10.
> 300 is a multiple of 100.

Show and Share

Share the products and patterns
you found with another pair of classmates.
How can you multiply by 10 and by 100
without using a calculator?
How can you multiply by multiples of 10 and
of 100 without using a calculator?

➤ You can use place value and patterns to multiply by 10 and by 100.

You know $3 \times 1 = 3$.

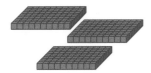

Use mental math to find 3×10 and 3×100.

3×1 ten $= 3$ tens
$\quad 3 \times 10 = 30$

3×1 hundred $= 3$ hundreds
$\quad\quad 3 \times 100 = 300$

➤ You can use basic multiplication facts and place value to multiply by multiples of 10 and of 100.

You know 2×4 ones $= 8$ ones
$\quad\quad\quad\quad 2 \times 4 = 8$

Use mental math to find 2×40 and 2×400.

2×4 tens $= 8$ tens
$\quad 2 \times 40 = 80$

2×4 hundreds $= 8$ hundreds
$\quad\quad 2 \times 400 = 800$

Use Base Ten Blocks when they help.

1. Use a basic fact and patterns to find each product.

   a) $6 \times 1 = \square$ **b)** $7 \times 3 = \square$ **c)** $4 \times 6 = \square$
   $\quad 6 \times 10 = \square$ $\quad 7 \times 30 = \square$ $\quad 4 \times 60 = \square$
   $\quad 6 \times 100 = \square$ $\quad 7 \times 300 = \square$ $\quad 4 \times 600 = \square$

2. Multiply.

   a) 3×10 **b)** 5×10 **c)** 7×10 **d)** 9×10

   e) 10×4 **f)** 10×1 **g)** 10×8 **h)** 10×0

3. Find each product.

   a) 4×100 **b)** 100×6 **c)** 9×100

   d) 100×1 **e)** 7×100 **f)** 100×0

4. There are 60 cards in one box.
   Caitlin bought 8 boxes.
   How many cards did Caitlin buy?
   How did you find out?

5. Multiply.

   a) 3×50 **b)** 4×70 **c)** 9×30

   d) 90×8 **e)** 20×6 **f)** 80×3

6. There are 200 cents in 1 toonie.
   Clay has 6 toonies.
   How many cents does Clay have?

7. Find each missing number.

   a) $10 \times \square = 60$ **b)** $\square \times 100 = 800$ **c)** $2 \times \square = 80$

   d) $4 \times \square = 200$ **e)** $20 \times \square = 180$ **f)** $\square \times 9 = 900$

   g) $\square \times 3 = 90$ **h)** $60 \times \square = 240$ **i)** $\square \times 7 = 70$

8. **a)** How many balloons are in 6 packages?
   b) How many candles are in 9 packages?
   c) How many napkins are in 7 packages?
   Show your work.

9. Make up some questions to show
how you can multiply by:
   a) multiples of 10
   b) multiples of 100
   Explain how your strategies work.

10. There are 50¢ in 1 roll of pennies.
How many cents are in 9 rolls of
pennies?

50 CENTS

   a) Write an equation you can solve to find out.
   b) Solve the equation. Answer the question.

11. One tower is made with 100 red blocks,
200 yellow blocks, and 200 white blocks.
   a) How many blocks of each colour
      would you need to make 4 towers?
   b) How many blocks would you need altogether?

12. Choose a number.
Choose a multiple of 10 or 100.
Write an equation you can use to find the
product of the number and the multiple.
Solve the equation.

Math Link

Measurement

To find the area of a rectangle, count the squares.
6 rows of 10 squares = 60 squares

To find the product of 6×10, use mental math.
$6 \times 10 = 60$

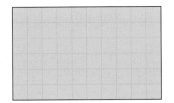

Reflect

Which patterns did you use when you multiplied by 10 and by 100?
Use words, pictures, or numbers to explain.

Estimating Products

Sometimes you do not need to know the exact amount.
You only need to know *about* how many or how much.
An estimate is close to the exact amount.

About 80 people are coming to the play. Do you think we have enough cookies?

It looks like there are about 20 cookies on each plate. That's about 100 cookies.

Explore

A Bombardier Challenger airplane holds 22 passengers.
About how many passengers will 8 of these planes hold?
Estimate to solve this problem.
Record your answer.

Show *and* Share

Share your strategies for estimating the number of passengers
with another pair of students.
Should your estimates be the same? Explain.

➤ A school bus holds 64 students.
About how many students can travel
on 7 school buses?

To estimate 7 × 64
↓

(Think:) 7 × 60 = 420

About 420 students can travel on
7 school buses.

I think of the closest
multiple of 10.
64 is close to 60.
I'll use 60 when I estimate.

➤ There are 87 pages in a book.
About how many pages are there
in 5 of these books?

To estimate 5 × 87
↓

(Think:) 5 × 90 = 450

There are about 450 pages in 5 books.

87 is close to 90.
I'll use 90 when I
estimate.

➤ A video costs $15.
About how much do 6 videos cost?

To estimate 6 × 15
↓

(Think:) 6 × 20 = 120

It costs about $120 to buy 6 videos.

15 is just as close
to 10 as it is to 20. I use
20 so I know I'll have
enough money.

Practice

1. Estimate each product.

    a) 3 × 21   **b)** 4 × 28   **c)** 5 × 35   **d)** 7 × 74

2. A can of soup costs 69¢.
About how much will 7 cans cost?

3. Estimate each product.
  a) 62 × 4 **b)** 57 × 8 **c)** 28 × 2 **d)** 43 × 9

4. A belt is 77 cm long.
About how long are 5 of these belts?
How do you know?

5. Estimate to find out which product is greater:
6 × 72 or 7 × 66

6. Kyle's mother drives 47 km to work 5 days a week.
About how far does she drive in 2 weeks?
Show your work.

7. The estimated answer to a multiplication question is 360.
What might the question be?
How do you know?

8. There are 35 students in each group.
There are 8 groups.
Ali estimates that there are about 240 students in all.
Jenny estimates that there are about 320 students.
Explain why the estimates are different.

Reflect

How do you choose the multiple of 10 when you estimate?
Use words and numbers to explain.

Using Models to Multiply

Explore

There are 24 eggs in a tray.
How many eggs are there in 6 trays?

Solve this problem.
Show your work.

Show *and* Share

Share your strategy with another pair of students.
Did you get the same answer?
If not, how can you find out who is correct?

Connect

There are 36 trees in each row in the new park. There are 4 rows.
How many trees have been planted?

4 rows of 36 trees $= 4 \times 36$
You can use models to multiply.

➤ Use Base Ten Blocks.
Arrange 4 rows of 3 tens and 6 ones.

Multiply the tens. $\longrightarrow$ 4×30 $\quad$ 4×6 $\longleftarrow$ Multiply the ones.
Add. $\qquad\qquad$ 120 $\quad + \quad$ 24 $\quad = 144$

$4 \times 36 = 144$ $\longleftarrow$ This **multiplication sentence** is an equation.

➤ Use grid paper to show an array for 4×36.

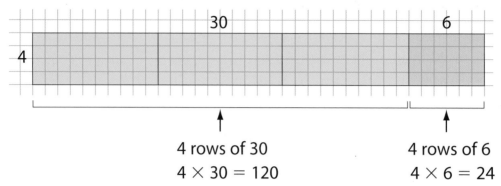

4 rows of 30   4 rows of 6
$4 \times 30 = 120$ $4 \times 6 = 24$

$$120 + 24 = 144$$

144 trees have been planted.

Practice

1. Write a multiplication sentence for each model.

a)

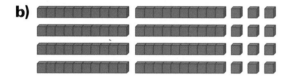

b)

2. On grid paper, draw an array to find each product.
   a) 5×41 b) 35×4 c) 6×25 d) 18×6

Use Base Ten Blocks or grid paper when they help.

3. Multiply.
   a) 23   b) 36   c) 62   d) 72   e) 47
      $\times 3$ $\times 2$ $\times 4$ $\times 6$ $\times 3$

4. Find each product.
   a) 5×61 b) 2×93 c) 45×4 d) 7×35 e) 19×5

5. Eva says to find 3×29, she would use mental math
   to find 3×30, then subtract 3.
   Is Eva correct? Explain.

6. Gita works at a garden centre.
She plants 15 seedlings in each row.
Gita plants 7 rows.
How many seedlings does Gita plant?
a) Write an equation you can solve to find out.
b) Solve the equation. Answer the question.

7. How much greater is 7 × 23 than 6 × 23? Explain.

8. Tom is buying candles for his
great grandmother's 90th birthday.
There are 24 candles in a box.
Tom buys 4 boxes of candles.
a) Will he have enough candles?
How do you know?
b) Will Tom have any candles left over?
How did you find out?
Show your work.

9. Write a story problem that can be solved by multiplying.
Solve your problem. Show your work.

10. Tara says that 4 × 36 is the same as
4 × 30 plus 4 × 6.
Do you agree? Explain your strategy.

11. A tray of petunias has 6 rows of 24 plants.
A tray of pansies has 8 rows of 16 plants.
Which tray has more plants?
Show your work.

At Home

Reflect

You have learned 2 models to
multiply. Which do you prefer?
Include an example of how
you used the model.

Ask relatives and friends what
strategies they use to multiply
two numbers such as 74 × 5.
Write about their strategies.

Strategies Toolkit

Explore

Marcus makes a "penny triangle."
He puts 1 penny in the 1st row, 2 pennies in the
2nd row, 3 pennies in the 3rd row, and so on.
How many pennies does Marcus need to make
a triangle with 8 rows?

Show *and* Share

Share your strategy with another pair of students.

Connect

There are 7 students in an inter-school math
competition. When they meet, each student
shakes hands with every other student.
How many handshakes will there be?

Strategies

- **Make a table.**
- **Use a model.**
- **Draw a picture.**
- **Solve a simpler problem.**
- **Work backward.**
- **Guess and test.**
- **Make an organized list.**
- **Use a pattern.**

What do you know?
- There are 7 students.
- When student A shakes hands with
  student B, that's *one* handshake.
- You have to find how many
  handshakes there will be in all.

Think of a strategy to help you solve the problem.
- You could **solve a simpler problem**.
- Count how many handshakes for 2 students, then for
  3 students, and so on. Look for a pattern in the answers.

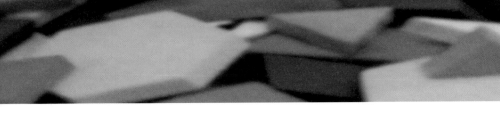

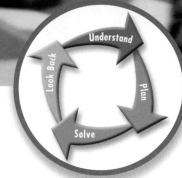

For 2 students, there is 1 handshake.
For 3 students, there are 3 handshakes.

Copy and continue this table.
How many handshakes are there for
7 students?

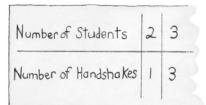

Number of Students	2	3
Number of Handshakes	1	3

How could you solve this problem another way?
Each student shakes hands with 6 other students.
Why is the total number of handshakes not 7×6?

Practice

Choose one of the

Strategies

1. Here is a pattern with Colour Tiles.
   Suppose the pattern continues.

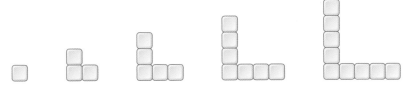

   a) How many tiles would be in the 7th figure?
   b) How many tiles are there in the first 7 figures?

2. How many squares can you see in this picture?
   Remember to count big squares as well as small squares.

Reflect

Explain how you used the strategy of solve a simpler
problem to solve one of the problems in this lesson.

Other Strategies for Multiplication

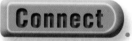

There are 56 balloons in each package.
Kim bought 7 packages for his carnival game.
How many balloons has he bought?

Use any materials that help.
Show your work.

Show *and* Share

Share your strategy for multiplying with another pair of classmates.
How do you know you have the correct answer?

Connect

The prizes for the Fun Fair have arrived.
Each package has 76 prizes.
There are 3 packages.
How many prizes are there?

Here are three ways to find out.

The total number of prizes is 76×3.

➤ Use Base Ten Blocks to model the problem.

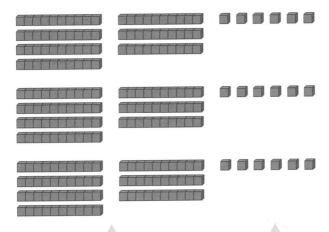

Multiply the tens.
3 × 70
210

Multiply the ones.
3 × 6
18

Add: 210 + 18 = 228
So, 76 × 3 = 228

➤ Write the number in expanded form: 76 = 70 + 6
Multiply the tens and multiply the ones.
Then add.

$$3 \times 76 = (3 \times 70) + (3 \times 6)$$
$$210 \quad + \quad 18 \quad = 228$$

76 × 3 = 210 + 18 = 228

I estimate to check my answer is reasonable.
76 is close to 80.
I know 3 × 80 = 240.
240 is close to the answer, 228.

➤ Break the number apart.

$$
\begin{array}{r}
76 \\
\times\ 3 \\
\end{array}
$$

Multiply the ones: 3 × 6 ⟶ 18
Multiply the tens: 3 × 70 ⟶ +210
Add.   228

So, 76 × 3 = 228

Use Base Ten Blocks when they help.

1. Find each missing number.
   a) $31 \times 7 = (30 \times 7) + (\square \times 7)$ **b)** $45 \times 8 = (40 \times \square) + (5 \times 8)$
   c) $66 \times 5 = (\square \times 5) + (6 \times 5)$ **d)** $86 \times 2 = (80 \times \square) + (6 \times \square)$

2. Find each product.

a) 29	**b)** 82	**c)** 66	**d)** 36	**e)** 41
$\times 5$	$\times 6$	$\times 3$	$\times 2$	$\times 8$

3. Multiply. Which strategies did you use?
   a) 19×5 **b)** 39×4 **c)** 55×3 **d)** 23×9 **e)** 78×2

4. Each class is 45 minutes long.
   The students have 3 classes after lunch.
   How many minutes are students in class after lunch?

5. Write a story problem that can be solved by
   multiplying a 2-digit number by a 1-digit number.
   Solve your problem.
   Show your work.

6. Noah says that 34×8 is the same as $240 + 32$.
   Do you agree?
   Use words, pictures, or numbers to explain.

7. Chris wrote this product to find 62×6:
   Explain each step of Chris' work.

$$\begin{array}{r} 62 \\ \times\ 6 \\ \hline 12 \\ +\ 360 \\ \hline 372 \end{array}$$

Reflect

Choose a *Practice* question. How can you
check your answer by using a different strategy?

Using Patterns to Multiply

You will need a copy of this multiplication chart.
Use patterns to complete the chart.

Show *and* Share

Show your completed chart to
another pair of students.
Talk about the patterns you used,
and the patterns in the chart.
Describe the pattern in the
products that have 11 as a factor.

x	1	2	3	4	5	6	7	8	9
10	10	20	30	40	50	60	70	80	90
11	11	22	33	44	55				
12	12	24	36	48	60				
13	13	26	39	52	65				
14	14	28	42	56	70				
15	15	30	45	60	75				
16	16	32	48	64	80				
17	17	34	51	68	85				
18	18	36	54	72	90				
19	19	38	57	76	95				
20									

You can use patterns and mental math to multiply.

➤ Multiply: 6×79

Think: 79 is 1 less than 80.
So, 6×79 is
6 less than 6×80.
$6 \times 80 = 480$
Subtract 6.
$480 - 6 = 474$
So, $6 \times 79 = 474$

➤ Multiply: 8×42

Think: 42 is 2 more than 40.
So, 8×42 is
8×40 plus 8×2.
$8 \times 40 = 320$
Add 8×2, or 16.
$320 + 16 = 336$
So, $8 \times 42 = 336$

Practice

1. Multiply. What patterns do you see?
   a) 2 × 99   **b)** 3 × 99   **c)** 4 × 99   **d)** 5 × 99   **e)** 6 × 99

2. Find each product.
   a) 43 × 8   **b)** 9 × 37   **c)** 5 × 72   **d)** 36 × 6   **e)** 7 × 17

3. Stickers cost 68¢ a sheet.
   How much money do you need for 6 sheets?

4. How can you tell what the ones digit of the product of
   53 × 7 will be without solving the whole problem?

5. These numbers are from one row of a multiplication chart:

117	126	135	144	153	162

   What number is being multiplied?
   How do you know? Show your work.

6. Copy and complete this multiplication chart.
   Use patterns to check.

×	60	61	62	63	64	65
2	120	122	124			
3		183	186	189		
4	240		248	252		
5	300		310	315		
6						
7						

Reflect

How can you use what you know about patterning
to help you multiply?

Multiplying a 3-Digit Number by a 1-Digit Number

Explore

Serena bought 2 packages of counters.
Each package contains 136 counters.
How many counters did Serena buy?

Use Base Ten Blocks to model the problem.
Write a multiplication fact for your model.
Record your work.
How can an estimate help you decide if your answer is reasonable?

Show *and* Share

Share your work with another pair of students.
How is multiplying a 3-digit number by a 1-digit number
like multiplying a 2-digit number by a 1-digit number?
How is it different?
What strategy did you use to estimate?

Connect

Mr. Martel arranged his class into 3 groups for an activity.
Each group needs a piece of string 145 cm long.
What length of string does Mr. Martel need?

The total length of string is 3 × 145 cm.

Here are three ways to multiply.

➤ Use Base Ten Blocks to model the problem.

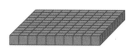
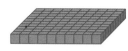
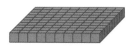

Multiply the hundreds.	Multiply the tens.	Multiply the ones.
3 × 100	3 × 40	3 × 5
300	120	15

Add: 300 + 120 + 15 = 435
So, 3 × 145 = 435

➤ Use expanded form.

Write 145 as 100 + 40 + 5.
Multiply, then add.

3 × 145 = (3 × 100) + (3 × 40) + (3 × 5)
            ↓   ↓   ↓
3 × 145 = 300   +   120   +   15 = 435

So, 3 × 145 = 435

> 145 is close to 150.
> 3 × 150 is 450. The answer,
> 435, is reasonable.

➤ Break a number apart to multiply.

$$\begin{array}{r} 145 \\ \times\ 3 \\ \end{array}$$

Multiply the ones: 3 × 5 ⟶ 15

Multiply the tens: 3 × 40 ⟶ 120

Multiply the hundreds: 3 × 100 ⟶ + 300

Add. 435

Mr. Martel needs 435 cm of string.

Practice

Use Base Ten Blocks when they help.

1. Find each missing number.
   a) 178 × 5 = (100 × □) + (70 × 5) + (8 × □)
   b) 523 × 4 = (500 × 4) + (20 × □) + (○ × 4)
   c) 234 × 5 = (□ × 5) + (30 × 5) + (4 × ○)
   d) 413 × 2 = (□ × 2) + (○ × 2) + (3 × 2)

2. Multiply. How do you know your answer is reasonable?
   a) 121   b) 216   c) 171   d) 412   e) 210
      × 3   × 4   × 5   × 3   × 6

3. Find each product.
   a) 3 × 492   b) 152 × 7   c) 5 × 215   d) 124 × 6   e) 2 × 198

4. A large box of crayons holds 128 crayons.
   How many crayons are in 4 large boxes?
   Estimate to check if your answer is reasonable.

5. Solve each equation.
   a) □ = 3 × 125   b) 256 × 4 = □   c) □ = 118 × 5

6. Write a story problem for each equation in question 5.

7. Write a story problem that can be solved by multiplying a 3-digit number by a 1-digit number.
Solve your problem.
Show your work.

8. Each seat on a roller coaster holds 3 people.
There are 42 seats.
The roller coaster completes 6 rides every hour.
Could 800 people ride the roller coaster in one hour? Explain.

9. Copy and complete each multiplication chart.
Explain your thinking.

a)

×	172	173	174	175
3	516		522	
4		692		700
5	860		870	
6		1038		1050

b)

×				
	600	603		
	800			812
			1010	1015
	1200		1212	1218

Reflect

How can you use what you know about multiplying a 2-digit number by a 1-digit number to multiply a 3-digit number by a 1-digit number?

Estimating Quotients

In a division fact, the answer is the **quotient**.

$12 \div 3 = 4$

          quotient

Explore

Students have collected 65 cans of food during
a food drive.
They will pack the cans into boxes before they
deliver them.
There are 9 boxes.
About how many cans will be in each box?
Record your answer.

Show *and* Share

Share your estimate with another pair of students.
Describe the strategies you used to estimate.

Connect

➤ A roll of ribbon is 82 m long.
Students plan to cut the ribbon into 9 equal pieces.
About how long will each piece be?

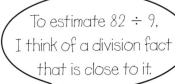

To estimate 82 ÷ 9,
I think of a division fact
that is close to it.

To estimate $82 \div 9$:

Think: 82 is close to 81.
                 81 is a multiple of 9.
                 $81 \div 9 = 9$
Each piece of ribbon will be about 9 m long.

➤ Nicole has $20.

She plans to buy small gifts for 7 friends.

About how much money will she spend on each gift?

Here are two ways to estimate.
- To estimate $20 \div 7$, use division.

 20 is close to 21.

21 is a multiple of 7.

$21 \div 7 = 3$

Each gift will cost about $3.

- To estimate $20 \div 7$, use multiplication.

 About how many groups of 7 are in 20?

7×3 is 21.

21 is close to 20.

$21 \div 7 = 3$

Each gift will cost about $3.

To estimate $20 \div 7$, I think about how many groups of 7 are in 20. I think of 7×3,

Practice

1. Write the division fact that helps you estimate each quotient.
   a) $27 \div 4$ **b)** $13 \div 2$ **c)** $39 \div 5$ **d)** $64 \div 7$ **e)** $43 \div 6$

2. Write the multiplication fact that helps you estimate each quotient.
   a) $36 \div 5$ **b)** $64 \div 9$ **c)** $53 \div 6$ **d)** $28 \div 9$ **e)** $19 \div 2$

3. Estimate each quotient.
   a) $14 \div 3$ **b)** $21 \div 4$ **c)** $29 \div 9$ **d)** $65 \div 8$ **e)** $19 \div 6$

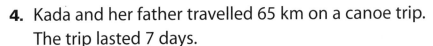

4. Kada and her father travelled 65 km on a canoe trip.
   The trip lasted 7 days.
   Kada travelled about the same distance each day.
   About how far did Kada travel each day?
   How do you know?

5. Estimate to find which quotient is greater:

      36 ÷ 7   or   50 ÷ 6

6. Forty-eight students go on a field trip.
   They are divided into 7 groups.
   About how many students are in each group?
   How do you know?

7. Is the quotient of 53 ÷ 6 greater than or less than 9?
   Explain your thinking.

8. Jilly has 65 stickers.
   She plans to share them among 9 friends.
   About how many will each friend get?
   Show your work.

9. Write a story problem you can solve
   by estimating the quotient.
   Solve your problem.
   Show your work.

10. Alona estimated 75 ÷ 9 as 8.
    Chung estimated 75 ÷ 9 as 7.
    Camille estimated 75 ÷ 9 as 9.
    Which personal strategy do you think each student used?
    Show your thinking.

Reflect

How does thinking of division and multiplication facts
help you to estimate?
Use words and numbers to explain.

Division with Remainders

Explore

Monica works in a market.
She arranges fruit baskets.
Monica has 41 oranges.
She puts 6 oranges in each basket.
How many baskets can Monica make up?
How many oranges are left over?
Use any materials that help. Show your work.

Show *and* Share

Share your answer with another pair of students.
Talk about the strategies you used.
How are they the same? How are they different?

Connect

➤ Monica has 25 apples.
She puts the same number of apples in each
of 4 baskets.
How many apples are there in each basket?

Share 25 apples equally among 4 baskets.
Divide: $25 \div 4$
Make an array, with 4 in each row.
There are 6 rows, with 1 left over.

Monica puts 6 apples in each basket.
There is 1 apple left over.
This is called a **remainder**.

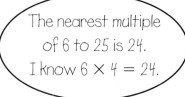

R stands for remainder.

You write: 25 ÷ 4 = 6 R1 ⟵ This is a **division sentence**.
You say: 25 divided by 4 is 6 remainder 1.

➤ Divide: 25 ÷ 6
Think about the division fact
that is closest to 25 ÷ 6.
You know that 24 ÷ 6 = 4.
So, 25 ÷ 6 = 4 R1

The nearest multiple of 6 to 25 is 24.
I know 6 × 4 = 24.

Practice

Use arrays when they help.

1. Write a division sentence for each array.

   a)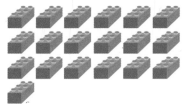

   b)

   c)

2. Divide. Draw an array to show how you got each answer.
   a) 17 ÷ 2 **b)** 28 ÷ 5 **c)** 24 ÷ 3 **d)** 20 ÷ 6

3. Caleb is putting his markers into packages.
   He has 43 markers.
   Each package holds 8 markers.
   a) How many packages will Caleb fill?
   b) How many markers will he have left over?

4. Which division statements have an answer greater than 6?
   How do you know?
   a) 50 ÷ 8   b) 45 ÷ 7   c) 76 ÷ 9   d) 13 ÷ 2
   e) 20 ÷ 4   f) 50 ÷ 6   g) 61 ÷ 8   h) 36 ÷ 5

5. Elizabeth takes 2 apples to school each day for her snack.
   She has 15 apples.
   How many days can Elizabeth take her snack to school?
   Show your work.

6. Divide.
   a) 14 ÷ 7   b) 15 ÷ 7   c) 16 ÷ 7   d) 17 ÷ 7
   e) 18 ÷ 7   f) 19 ÷ 7   g) 20 ÷ 7   h) 21 ÷ 7
   What is the greatest possible remainder when you divide by 7?
   How do you know it is the greatest?

7. Write a story problem that has a remainder
   when you divide to solve the problem.
   Solve the problem.

8. Amina solves a division problem this way: 21 ÷ 4 = 5 R1
   Tyler solves the problem this way: 21 ÷ 4 = 4 R5
   Who is correct? How do you know?
   Show your work.

9. Bottles are packaged 6 to a carton.
   Every bottle must be in a carton.
   There are 32 bottles to be packaged.
   a) How many cartons are needed?
   b) Does the number of cartons change if there are 35 bottles
      instead of 32? Explain.

Reflect

When you solve a division problem,
what strategies can you use?
Use examples to show your ideas.

Using Base Ten Blocks to Divide

Explore

Felipe has 76 books.
He divides them equally among 4 boxes.
How many books are in each box?
Show your work.

Suppose Felipe had 78 books.
Could he divide them equally among 4 boxes?
How do you know?

Show *and* Share

Share your answers with those of another
pair of students.
What strategies did you use to solve the problem?

Connect

➤ Divide: 36 ÷ 3
   Use Base Ten Blocks to show 36.

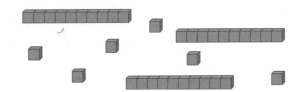

Divide the blocks into 3 equal groups.

In each group, there is 1 ten rod and 2 unit cubes.
So, there are 12 in each group.

36 ÷ 3 = 12

➤ Divide: 57 ÷ 4

Use Base Ten Blocks to show 57.

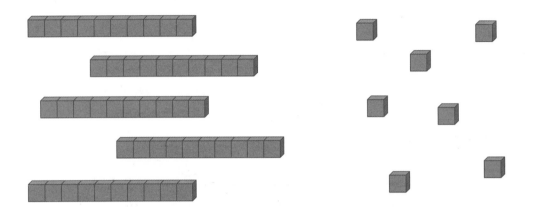

Divide the blocks into 4 equal groups.

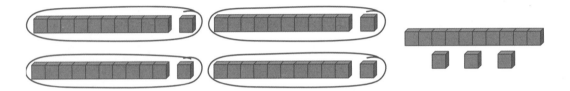

There is 1 ten rod and 1 unit cube in each group.
There is 1 ten rod and 3 unit cubes left over.
Trade the ten rod for 10 unit cubes.

There are 13 unit cubes.
Divide these cubes among the 4 equal groups.

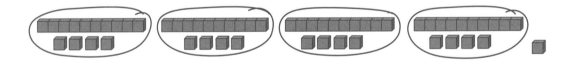

There is 1 ten rod and 4 unit cubes in each group.
There is 1 unit cube left over.

So, 57 ÷ 4 = 14 R1

Use Base Ten Blocks when they help.

1. Divide.
   a) 69 ÷ 3   b) 68 ÷ 4   c) 87 ÷ 2   d) 64 ÷ 4   e) 75 ÷ 6

2. Aidan is collecting eggs at a farm. He puts the eggs in cartons.
   Each carton holds 6 eggs. Aidan collects 34 eggs.
   How many cartons does he need?

3. Divide. Draw a picture of the blocks you used to get one answer.
   a) 93 ÷ 3   b) 49 ÷ 4   c) 96 ÷ 8   d) 56 ÷ 5   e) 91 ÷ 7

4. Write a story problem that can be solved using 78 ÷ 6.
   Solve the problem.
   Show your work.

5. Divide.
   a) 40 ÷ 2   b) 41 ÷ 2   c) 42 ÷ 2   d) 43 ÷ 2
   e) 44 ÷ 2   f) 45 ÷ 2   g) 46 ÷ 2   h) 47 ÷ 2
   How can you tell *before* you divide by 2 if there will be a remainder?

6. Divide.
   a) 40 ÷ 5   b) 42 ÷ 5   c) 45 ÷ 5   d) 46 ÷ 5
   e) 50 ÷ 5   f) 54 ÷ 5   g) 55 ÷ 5   h) 57 ÷ 5
   How can you tell *before* you divide by 5 if there will be a remainder?

7. Chin-Tan found 52 action figures for his yard sale.
   He wants to put them in more than 1 box,
   but fewer than 5 boxes. Each box will have
   the same number of figures.
   How many boxes can Chin-Tan use? Explain.

Reflect

Use what you have learned about remainders.
Which numbers have no remainder when they are
divided by 2? By 5? How do you know?

Another Strategy for Division

There are 63 trees.
They are to be planted in 4 equal rows.
How many trees will there be in each row?
Do you think there will be any trees left over?
How do you know?

Show *and* Share

Talk with another pair of students about
the strategy you used to solve this problem.

Connect

There are 76 plants.
They are to be planted in 3 gardens.
Each garden will have the same number of plants.
How many plants will there be in each garden?

Divide: $76 \div 3$
Use Base Ten Blocks to show 76.

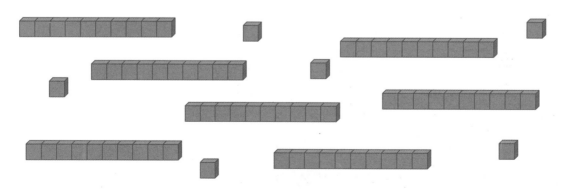

Arrange the 7 rods in 3 equal rows.

You see:

You think: You write:

7 rods ÷ 3
is 2 rods each
with 1 rod left over.

$$\begin{array}{r} 2 \\ 3\overline{)7\,^16} \end{array}$$

Trade 1 ten rod for 10 ones.
You have 16 unit cubes.

Share these 16 cubes equally among the 3 rows.

You see:

You think: You write:

16 cubes ÷ 3 is
5 cubes each
with 1 cube left over.

$$\begin{array}{r} 25\ R1 \\ 3\overline{)7\,^16} \end{array}$$

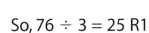

So, 76 ÷ 3 = 25 R1
There will be 25 plants in each garden.
There will be 1 plant left over.

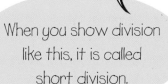

When you show division like this, it is called short division.

Practice

Use Base Ten Blocks when they help.

1. Find 3 division statements that have an answer greater than 11.
   a) 27 ÷ 2 **b)** 47 ÷ 4 **c)** 61 ÷ 6 **d)** 84 ÷ 8
   e) 52 ÷ 5 **f)** 46 ÷ 3 **g)** 99 ÷ 9 **h)** 73 ÷ 7

2. Victoria shares 49 crayons among 8 students.
   How many crayons does each student get?

3. Divide.
  a) 56 ÷ 6   **b)** 29 ÷ 9   **c)** 47 ÷ 7   **d)** 74 ÷ 4
  e) 92 ÷ 2   **f)** 83 ÷ 3   **g)** 38 ÷ 8   **h)** 65 ÷ 5

4. Emma is collecting a series of books.
Each book costs $6.
How many books can Emma buy with $53?

5. Trenton has to feed 8 cats.
He has 45 large cans of cat food.
Each large can feeds 2 cats per day.
How many days of cat food does Trenton have?
Show your work.

6. Divide.
  a) 36 ÷ 3   **b)** 38 ÷ 3   **c)** 39 ÷ 3   **d)** 40 ÷ 3
  e) 42 ÷ 3   **f)** 43 ÷ 3   **g)** 45 ÷ 3   **h)** 46 ÷ 3
  How can you tell *before* you divide by 3 if there will be a remainder?

7. Suppose you have 60 straws.
How many of each shape could you make?
  a) triangles
  b) squares
  c) pentagons
  d) hexagons

8. How many different 2-digit numbers can you find that
have remainder 2 when each is divided by 6?
List the numbers.
What strategy did you use to find them?

Reflect

You have used different strategies to divide.
Which strategy do you prefer?
Use words, numbers, or pictures to explain.

Less Is More

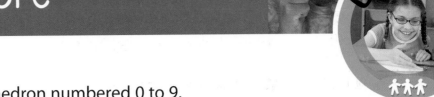

You will need a decahedron numbered 0 to 9.

The object of the game is to make a division sentence with:

- the least quotient, *and*
- the least remainder

➤ Your teacher will give you copies of this division frame.

☐ ☐ ÷ ☐ = _____

> A decahedron is an object with 10 congruent faces.

➤ Players take turns to roll the decahedron.

➤ On your turn, record the number that turns up in any square of the frame.
Once a number is written, you may not move it.

➤ Continue until each player has filled her or his frame.

➤ Each player finds the quotient for her or his frame.
Check each other's work.
Each player with a correct answer scores 1 point.
The player with the least remainder scores 1 point.
The player with the least quotient scores 1 point.
The first player to score 6 points wins.

LESSON

1

1. Find each product.
 a) 6×700 b) 900×8 c) 5×60 d) 80×4
 e) 200×5 f) 3×70 g) 7×400 h) 90×2

2. Find each missing number.
 a) $5 \times \square = 300$ b) $20 \times \square = 140$ c) $\square \times 6 = 600$
 d) $\square \times 7 = 210$ e) $40 \times \square = 240$ f) $\square \times 9 = 90$

3. A radio station gives away a $300 prize every day for a week.
 How much will the radio station have given away by the end of the week?
 Show your work.

WE'VE GOT ANOTHER WINNER!

2

4. Estimate each product.
 a) 5×31 b) 7×63 c) 8×56 d) 4×69

3
5

5. There are 6 rows of chairs set up for the concert.
 In each row, there are 45 chairs.
 How many chairs are there?
 Show your work.

6. Multiply. What strategies did you use?
 a)   29   b)   73   c)   34   d)   95
     $\times\,2$ $\times\,3$ $\times\,6$ $\times\,4$

6

7. Copy and complete this multiplication chart.
 Explain how you could use patterns to do this.

×	85	86	87	88	89
5	425	430	435		
6	510	516			
7	595				
8					
9					

8. Identify the errors in this multiplication chart.
How did you identify each error?
Correct each error.

×	25	30	35	40	45
2	50	60	75	80	90
4	100	110	140	160	180
6	150	180	210	240	260
8	180	240	280	320	360

9. Multiply.

a) 178
   × 6

b) 319
   × 3

c) 164
   × 2

d) 462
   × 5

10. Estimate each quotient.
Which facts helped you estimate?

a) 32 ÷ 6   b) 65 ÷ 8
c) 26 ÷ 9   d) 43 ÷ 7

11. Divide. What strategies did you use?

a) 42 ÷ 3   b) 52 ÷ 4
c) 65 ÷ 5   d) 78 ÷ 6
e) 91 ÷ 7   f) 88 ÷ 8
g) 99 ÷ 9   h) 34 ÷ 2

12. A series on TV runs for 25 hours.
One videotape can record 4 hours.
Write a word problem using these data.
Solve the problem.
Show your work.

13. Divide.

a) 76 ÷ 5   b) 65 ÷ 3
c) 21 ÷ 2   d) 32 ÷ 6
e) 98 ÷ 7   f) 54 ÷ 8
g) 87 ÷ 9   h) 43 ÷ 4

14. Divide. What patterns do you see?

a) 99 ÷ 3   b) 98 ÷ 3
c) 97 ÷ 3   d) 96 ÷ 3
e) 95 ÷ 3   f) 94 ÷ 3

UNIT

8 Learning Goals

☑ use personal strategies to multiply
☑ estimate products
☑ use models and arrays to multiply and divide
☑ multiply a 2-digit and a 3-digit number by a 1-digit number
☑ estimate quotients
☑ divide a 2-digit number by a 1-digit number
☑ use personal strategies to divide
☑ relate multiplication and division
☑ identify patterns in multiplication and division

1. Jean works at a garden centre.
   He has an order for 72 petunias.
   The petunias are grown in boxes of 4 or 9.
   How many boxes of each size does Jean need?
   Can he deliver the order in more than one way?
   Explain.

2. The boxes of petunias fit on trays.
   One tray holds 6 boxes of 4 petunias or
   3 boxes of 9 petunias.
   How many trays are needed for an order of 75 petunias?

3. May-Lin is replanting trees.
She has 80 trees.
May-Lin will plant them in equal rows.
How many different ways can she do this?
Show each way as a multiplication fact,
then a division fact.

4. The garden centre sells small plastic pots
to grow seedlings.
The pots are sold in packages of 30 or 50.
One package of 30 pots costs $7.
One package of 50 pots costs $9.
A customer wants 180 pots.
What is the cheapest way she can buy the pots?

Reflect on Your Learning

Describe the strategies you use to multiply and divide.
Which strategies do you need to practise?
Give an example for each strategy.

Investigation

Circle Patterns

You will need a ruler and two copies of the sheet of circle diagrams below.

Part 1

➤ Write the first 12 multiples of 2.
➤ List the ones digit of each multiple.
  What do you notice?
➤ Use a circle diagram.
  On the circle, find the first number on your list.
  Draw a line from this number to the second number on the list,
  then from the second to the third, and so on.

Part 2

➤ Repeat *Part 1* for multiples of 8.
➤ What do you notice about your diagrams?
➤ Write about the diagrams.

Part 3

➤ Repeat *Parts 1* and *2* for multiples of 3 and multiples of 7.
➤ How are your diagrams similar to those for multiples of 2
  and multiples of 8?
  How are they different?

Part 4

➤ Predict the patterns you will make for multiples of 4 and 6.
  Check your predictions.

Display Your Work

Make a poster display of your
number patterns and circle diagrams.

Take It Further

➤ Predict other pairs of numbers whose
  multiples will produce the same diagrams.
  Explain your thinking.
➤ Check your predictions.

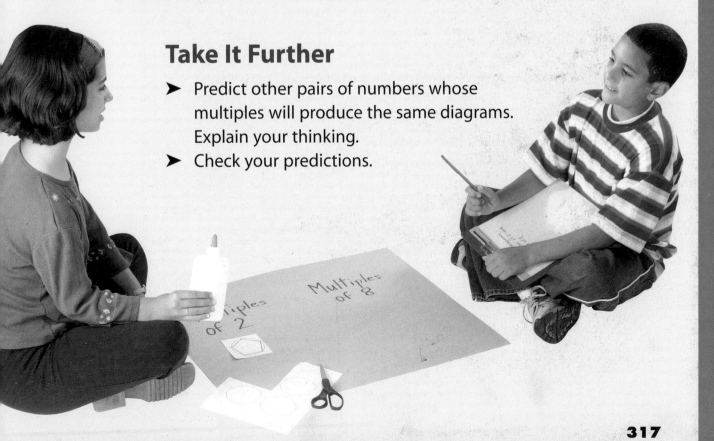

UNIT

1

1. Here is a pattern made with Colour Tiles.
The side length of each square is 1 unit.

         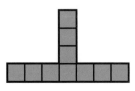

Figure 1   Figure 2   Figure 3

a) Draw the next 3 figures on grid paper.
b) Copy and complete this table for the first 6 figures.

Figure	Perimeter (units)
1	

c) Write a pattern rule for the perimeters.
d) Predict the perimeter of the 10th figure.
e) Will any figure have a perimeter of 50 units?
How did you find out?

2. Say what each equation means. Then solve each equation.
a) $\nabla + 6 = 15$ **b)** $56 = 7 \times \square$ **c)** $12 = \bigcirc - 20$ **d)** $\heartsuit \div 6 = 9$

3. Write a story problem that could be solved
using each equation in question 2.

2

4. A school concert had 3 performances.
This table shows how many people came each day.

Thursday	Friday	Saturday
1357	3408	2991

a) How many people went to the concert altogether?
b) How many more people went on Friday than on Thursday?
Than on Saturday?

5. a) Copy this Carroll diagram.
Sort these numbers
in the Carroll diagram:
12, 19, 24, 27, 28,
30, 37, 40, 52, 71

	Is divisible by 6	Is not divisible by 6
Is odd		
Is not odd		

b) Why is one box empty?
Write another number in each of the other 3 boxes.

c) Use the numbers from parts a and b.
Sort the numbers in a Venn diagram.
Use the attributes "Odd" and "Divisible by 6."

d) Do the Carroll diagram and Venn diagram
show the same information?
Explain how you know.

6. Explain the meaning of each digit in the number 8888.

7. Write each number in standard form.
a) $5000 + 300 + 20 + 1$ **b)** $6000 + 50$

8. Draw an array to illustrate your answer to each question below.
a) What is the quotient when you divide any number by 1?
b) What is the quotient when you divide any number by itself?

9. Suppose you know that $3 \times 4 = 12$.
What other multiplication facts can you find?
Explain how you found each fact.

10. Write each date in metric notation.
a) November 5th, 1999 **b)** July 7th, 2005 **c)** April 16th, 1950

11. Each date is written in metric notation.
Write each date using words and numbers.
a) 1998 03 14 **b)** 2007 10 10 **c)** 1997 06 03

12. a) Use a benchmark for 1 cm². Tell how you could estimate the area of the cover of your math book.

b) Use a benchmark for 1 m². Tell how you could estimate the area of your classroom floor.

5

13. Draw a picture for each decimal. Write the decimal as a fraction.
a) 0.2   **b)** 0.02   **c)** 0.5   **d)** 0.05   **e)** 0.50

14. Draw a picture for each fraction. Write the fraction as a decimal.
a) $\frac{3}{10}$ **b)** $\frac{4}{100}$ **c)** $\frac{9}{10}$ **d)** $\frac{40}{100}$ **e)** $\frac{89}{100}$

15. When is $\frac{1}{2}$ of one set not equal to $\frac{1}{2}$ of another set?

16. Samya bought juice for $1.60 and fruit salad for $2.49. How much change did she get from a $5 bill?

6

17. Find 4 triangular prisms in the classroom. How are the prisms alike?

18. Use modelling clay. Make a triangular prism and a rectangular prism. How are the prisms alike? How are they different?

19. Use Pattern Blocks. Make a design that is symmetrical. Copy the design on dot paper. How do you know it is symmetrical?

7

20. The tally chart shows the favourite animals for students in Grades 4 and 5.

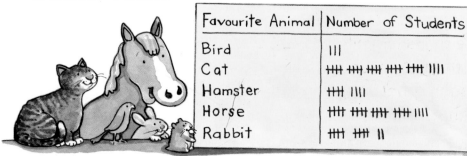

Favourite Animal	Number of Students
Bird	|||
Cat	++++ ++++ ++++ ++++ ++++ ||||
Hamster	++++ ||||
Horse	++++ ++++ ++++ ++++ ||||
Rabbit	++++ ++++ ||

320

a) Draw a bar graph to display these data.
What scale did you use? Explain your choice.
b) Draw a pictograph to display the data.
What key did you use? Explain your choice.
c) Which animal is twice as popular as the rabbit?
d) How many more students chose a cat than a hamster?
e) Write a question you can answer using the bar graph or pictograph. Answer your question.

21. Find each product. What strategies did you use?

a) 6×7	**b)** 8×3	**c)** 5×9	**d)** 4×5
6×70	8×30	5×90	4×50
6×700	8×300	5×900	4×500

22. There are 8 nickels and 7 dimes in a change purse.
How many cents is that?

23. Estimate each quotient.
Which quotients are greater than 9?

a) $37 \div 3$	**b)** $46 \div 9$	**c)** $58 \div 5$	**d)** $63 \div 8$

24. Estimate each product.
Which strategy did you use each time?

a) 5×52	**b)** 68×6	**c)** 4×44	**d)** 9×32

25. Multiply.
How do you know your answer is reasonable?

a) 2×198	**b)** 4×136	**c)** 333×3	**d)** 164×5

26. There are 85 counters.
They are to be shared equally among 6 students.
Each student needs 14 counters.
Are there enough counters?
How do you know?

Illustrated Glossary

A.M.: A time between midnight and just before noon.

Addition fact: 3 + 4 = 7 is an addition fact. The sum is 7. See also **Related facts**.

Analog clock: A clock that shows time by hands moving around a dial.

This clock shows 9 o'clock.

Area: The number of congruent units that cover a shape. We also measure area in square units, such as square centimetres and square metres.

The unit is 1 green block. The area of the yellow block is 6 green blocks.

Array: A set of objects arranged in equal rows.

Attribute: A way to describe a shape or an object; for example, number of sides, number of vertices.

Axis (plural: axes): A number line along the edge of a graph. We label each axis of a graph to tell what data it displays. See **Bar graph**.

Bar graph: Displays data by using bars of equal width on a grid. The bars may be vertical or horizontal.

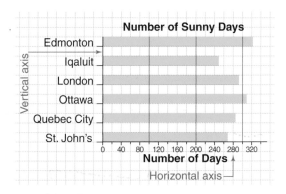

Base: The face that names an object.

Base Ten Blocks: Blocks used to model whole numbers. Here is one way to model 2158:

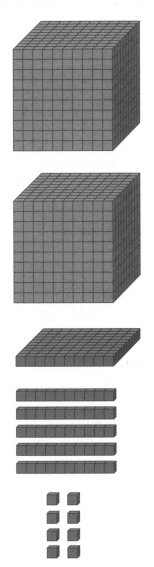

Base-ten name: See **Place value**.

Capacity: A measure of how much a container holds.

Centimetre: A unit to measure length. We write one centimetre as 1 cm. One centimetre is one-hundredth of a metre.

Century: A unit of time equal to 100 years.

Compare: To look at how items are alike as well as different.

Cone: An object with a circular base, a curved surface, and a vertex.

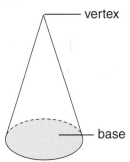

Congruent figures: Two shapes that match exactly. The shapes may face different ways.

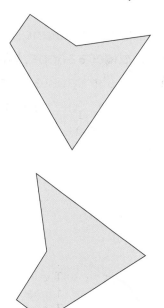

Core: See **Repeating pattern**.

Cube: An object with 6 faces that are congruent squares. Two faces meet at an edge. Three edges meet at a vertex.

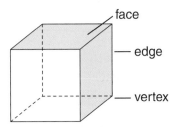

Cylinder: An object with 2 congruent circular bases joined by a curved surface.

Data: Facts or information collected to learn about people or things.

Decade: A unit of time equal to 10 years.

Decagon: A polygon with 10 sides.

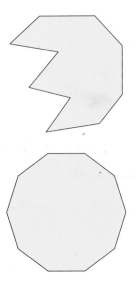

Decimal: A way to write a fraction; the fraction $\frac{2}{10}$ can be written as the decimal 0.2.

Decimal point: Separates the whole number part and the fraction part in a decimal. We read the decimal point as "and." We say 0.2 as "zero **and** two-tenths."

Denominator: The part of a fraction that tells how many equal parts are in one whole.
The denominator is the bottom number in a fraction.

Diagonal: A line segment that joins opposite corners or vertices of a shape.

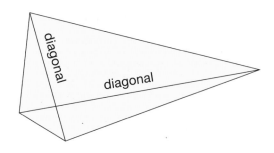

Difference: The result of a subtraction. The difference of 5 and 2 is 3; or 5 − 2 = 3.

Digit: See **Place value.**

Digital clock: A clock with numbers and no hands.

This clock shows 9 o'clock.

Divide: To separate into equal parts.

Divisible: A number is divisible by 5 if that number of counters can be divided into 5 equal groups.

15 can be divided into 3 groups of 5. We say: 15 is divisible by 5.

Division sentence: $6 \div 3 = 2$ is a division sentence. We say: 6 divided by 3 equals 2.

Doubles: **1.** The addition of 2 numbers that are the same. Doubles have an even sum. **2.** Rolling two of the same number when using number cubes.

Edge: See **Cube**.

Elapsed time: The amount of time that passes from the start to the end of an event. The elapsed time between when you eat lunch and the end of school is about 3 hours.

Equation: Uses the = symbol to show two things that represent the same amount. $5 + 2 = 7$ is an equation.

Estimate: Close to an amount or value, but not exact.

Face: See **Cube**.

Factor: Numbers that are multiplied to get a product. In the multiplication fact, $3 \times 7 = 21$, the factors of 21 are 3 and 7.

Fractions: Common fractions: one-half ($\frac{1}{2}$), one-third ($\frac{1}{3}$), one-fourth or one-quarter ($\frac{1}{4}$), one-fifth ($\frac{1}{5}$), one-sixth ($\frac{1}{6}$), one-eighth ($\frac{1}{8}$), and one-tenth ($\frac{1}{10}$).

Friendly number: a number that is close to a given number but easier to add, subtract, multiply, or divide. A friendly number for 198 is 200. A friendly number for 403 is 400.

Gram: A unit to measure mass. We write one gram as 1 g. 1000 g = 1 kg

Half hour: One-half of an hour, or 30 minutes.

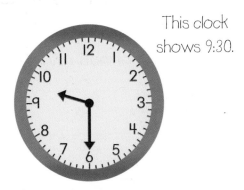

This clock shows 9:30.

Height: The measurement from top to bottom of an object.

Hexagon: A polygon with 6 sides.

Horizontal: To the left or right.

Hundredth: A fraction that is one part of a whole when it is divided into 100 equal parts. We write one hundredth as $\frac{1}{100}$ or 0.01.

Key: See **Pictograph**.

Kilogram: A unit to measure mass. We write one kilogram as 1 kg. 1 kg = 1000 g

Kilometre: A unit to measure long distances. We write one kilometre as 1 km. 1 km = 1000 m

Kite: A shape with 4 sides where two pairs of adjacent sides are equal.

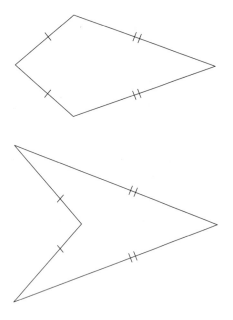

Length: The measurement from end to end; how long something is. See **Shape**.

Line of symmetry: Divides a shape into two congruent parts. If we fold the shape along its line of symmetry, the parts match.

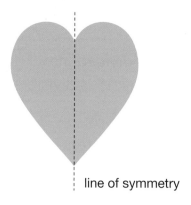

line of symmetry

Mass: Measures how much matter is in an object. We measure mass in grams or kilograms.

Metre: A unit to measure length. We write one metre as 1 m. 1 m = 100 cm

Multiple: Start at a number then count on by that number to get the multiples of that number. Start at 3 and count on by 3 to get the multiples of 3: 3, 6, 9, 12, 15, …

Multiplication fact: A sentence that relates factors to a product. $3 \times 7 = 21$ is a multiplication fact.

Net: A cutout that can be folded to make a prism.

Object: Has length, width, and height. Objects have faces, edges, vertices, and bases. We name some objects by the number and shape of their bases.

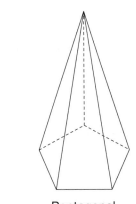

Pentagonal pyramid

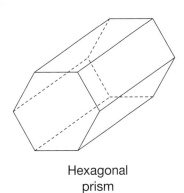

Hexagonal prism

Number line: Has equally spaced numbers marked in order.

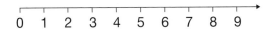

0 1 2 3 4 5 6 7 8 9

Numerator: The part of a fraction that tells how many equal parts to count. The numerator is the top number in a fraction.
In the fraction $\frac{2}{3}$, the numerator is 2, and we count 2 thirds of the whole.

Octagon: A polygon with 8 sides.

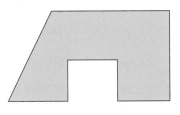

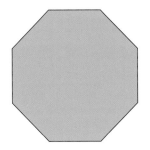

Ordinal number: A number that shows order or position: 2nd, 4th, 25th.

P.M.: A time between noon and just before midnight.

Pattern rule: Describes a pattern or how to make a pattern. To make the pattern 1, 2, 4, 8, 16, … use the pattern rule: Start with 1, then multiply by 2 each time.

Perimeter: The distance around a shape.
We can find the perimeter by measuring and adding side lengths. The perimeter of this rectangle is: 2 cm + 4 cm + 2 cm + 4 cm = 12 cm

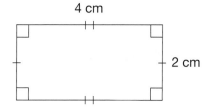

4 cm

2 cm

Pictograph: Uses symbols to display data. Each symbol can represent more than one object. A key tells what each symbol represents.

Videos Rented by Ali

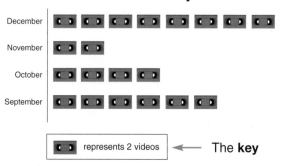

December
November
October
September

represents 2 videos ← The **key**

Place value: The value of each digit in a number depends on its place in the number.

Hundreds	Tens	Ones
4	9	7

The value of this digit is 4 hundreds, or 400.

The value of this digit is 9 tens, or 90.

The value of this digit is 7 ones, or 7.

The base-ten name is 4 hundreds 9 tens 7 ones.

Polygon: A shape with three or more straight sides. We name a polygon by the number of its sides. For example, a five-sided polygon is a pentagon.

Prism: An object with 2 congruent bases. The shape of the bases gives the name of the prism.

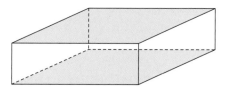

Rectangular prism

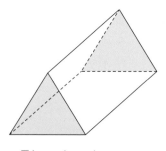

Triangular prism

Product: The result of a multiplication.
The product of 5 and 2 is 10; or $5 \times 2 = 10$.

Proper fraction: Describes an amount less than one. A proper fraction has a numerator that is less than its denominator.
$\frac{5}{7}$ is a proper fraction.

Pyramid: An object with 1 base and triangular faces. The shape of the base gives the name of the pyramid.

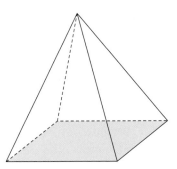

Rectangular pyramid

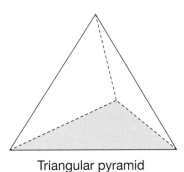

Triangular pyramid

Quadrilateral: A polygon with 4 sides.

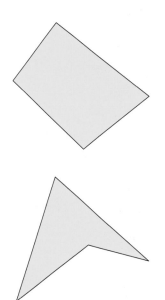

Quarter hour: One-fourth of an hour, or 15 minutes.

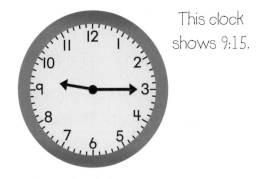

This clock shows 9:15.

Rectangle: A shape with 4 sides, where 2 pairs of opposite sides are equal and each angle is a right angle.

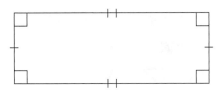

Rectangular prism: See **Prism**.

Regular polygon: A polygon with all sides equal and all angles equal. Here is a regular triangle.

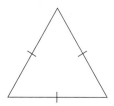

Related facts: Addition and subtraction have related facts. $2 + 3 = 5$ is related to $5 - 3 = 2$ and $5 - 2 = 3$. Multiplication and division have related facts. $5 \times 6 = 30$ is related to $30 \div 5 = 6$ and $30 \div 6 = 5$.

Remainder: What is left over when one number does not divide exactly into another number. For example, $12 \div 5 = 2$ R2

Repeating pattern: A pattern with a core that repeats. The core is the smallest part of a repeating pattern. The repeating number pattern: 1, 8, 2, 1, 8, 2, 1, 8, 2,… has a core of 1, 8, 2.

Right angle: Two lines that meet in a square corner make a right angle.

Scale: The number of items each square on a bar graph represents.

Shape: A geometric drawing or diagram.

Circle

Square

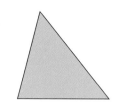

Triangle

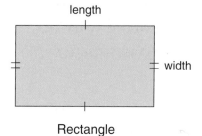

Rectangle

Skeleton: The frame of an object that shows the edges and vertices of the object. This is the skeleton of a pentagonal prism:

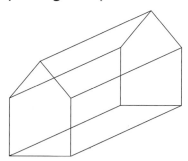

Sphere: An object shaped like a ball.

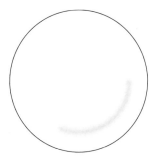

Square: A shape with 4 equal sides and each angle is a right angle.

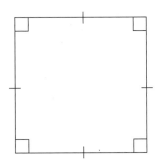

Square centimetre: A unit of area that is a square with 1-cm sides. We write one square centimetre as 1 cm².

Square metre: A unit of area that is a square with 1-m sides. We write one square metre as 1 m².

Square unit: Area is measured in square units.

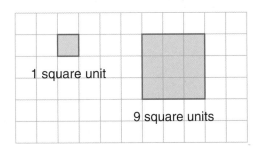

Standard form: Numbers written using digits. For example, 9372

Subtraction facts: 11 − 7 = 4 and 11 − 4 = 7 are subtraction facts. See also **Related facts**.

Sum: The result of an addition. The sum of 5 and 2 is 7; or 5 + 2 = 7.

Survey: Used to collect data. You can survey your classmates by asking them which is their favourite ice cream flavour.

Tally chart: A chart on which a count is kept.

Favourite Colours of Grade 4 Students

Black	⳾⳾⳾⳾⳾
Red	⳾⳾⳾⳾⳾ ⳾⳾⳾⳾⳾
Orange	⳾⳾⳾⳾⳾ ⳾⳾
Green	⳾⳾⳾⳾⳾ ⳾⳾⳾

Tenth: A fraction that is one part of a whole when it is divided into ten equal parts. We write one-tenth as $\frac{1}{10}$ or 0.1.

Term: One number in a number pattern. For example, the number 4 is the third term in the pattern 1, 2, 4, 8, 16, …

Triangle: A polygon with 3 sides.

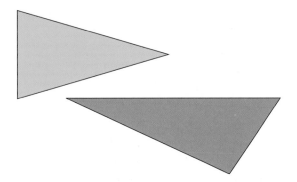

Triangular prism: See **Prism**.

Triangular pyramid: See **Pyramid**.

Unit: A standard amount used for measuring.

Venn diagram: A diagram that is used to sort items.

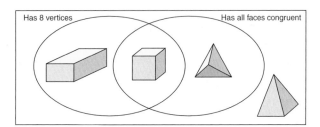

Vertex (plural: vertices): 1. A point where two sides of a shape meet. **2.** A point where two or more edges of an object meet.

Vertical: Straight up and down.

Width: The distance across something; how wide it is. See **Shape**.

Index

Acknowledgments

Pearson Education Canada wishes to thank the Royal Canadian Mint for the illustrative use of Canadian coins in this textbook. In addition, the publisher wishes to thank the following sources for photographs, illustrations, and other materials used in this book. Care has been taken to determine and locate ownership of copyright material in this text. We will gladly receive information enabling us to rectify any errors or omissions in credits.

Photography

Cover: Mark Hamblin/age fotostock
pp. 2-3 Ray Boudreau; p. 4 (left) Bryan and Cherry Alexander Photography/Alamy; p. 4 (right) Stone/Frank Siteman; p. 5 (left) Stone/Rick Rusing; p. 5 (right) Taxi/Hans Neleman; p. 9 Ray Boudreau; p. 10 Ray Boudreau; p. 14 Ian Crysler; p. 21 Jeff Greenberg/PhotoEdit, Inc.; p. 25 Ray Boudreau; pp. 30-31 Ray Boudreau; p. 32 Frans Lanting/CORBIS; p. 33 Besndeo/Getty Images; p. 34 (top) A. Ramey/PhotoEdit, Inc.; p. 34 (bottom) Ray Boudreau; p. 37 Ray Boudreau; p. 42 Ian Crysler; p. 54 Stockbyte; p. 55 Used by permission of Scott Suko; p. 65 Ian Crysler; p. 67 Corel; p. 68 Ray Boudreau; p. 70 Richard Lord/PhotoEdit, Inc.; p. 73 (left) Walter Bibikow/Taxi/Getty Images; p. 73 (centre) Super Stock/MaXx Images; p. 73 (right) Cn Boon/Alamy; p. 78 Used with permission and thanks to the Calgary Zoo; p. 78 (insets) Used by permission of the Calgary Zoo; p. 80 Mary Kate Denny/PhotoEdit, Inc.; p. 93 Ray Boudreau; p. 94 Ray Boudreau; p. 97 Ray Boudreau; p. 101 Ray Boudreau; p. 105 Shaun Best/Canapress; p. 107 Taxi/Jeff Kaufman; p. 119 Rudi Von Briel/PhotoEdit, Inc.; p. 126 Tom Hanson/Canadian Press CP; p. 128 Ian Crysler; p. 135 Ray Boudreau; p. 136 Ian Crysler; p. 137 James Gritz/Photodisc; p. 139 Ray Boudreau; p. 141 Ray Boudreau; p. 143 Ray Boudreau; p. 144 Ken Straiton/firstlight.ca; p. 146 Ray Boudreau; p. 150 Ray Boudreau; p. 153 Ray Boudreau; p. 154 Ian Crysler; p. 159 (top) Ray Boudreau; p. 159 (bottom) Ian Crysler; p. 168 Ray Boudreau; p. 169 (left) Lawrence Migdale/Stone; p. 169 (right) Michael Newman/PhotoEdit, Inc.; pp. 170-171 Ray Boudreau; p. 174 Ray Boudreau; p. 177 Tom Carter/Index Stock Imagery; p. 180 Ray Boudreau; p. 183 Ray Boudreau; p. 186 Ian Crysler; p. 188 Ray Boudreau; p. 190 Ray Boudreau; p. 192 Ray Boudreau; p. 193 Ian Crysler; p. 197 Ray Boudreau; p. 200 Ray

Illustrations